Lecture Notes in Mathematics

Volume 2337

This series reports on new developments in all areas of mathematics and their applications - quickly, informally and at a high level. Mathematical texts analysing new developments in modelling and numerical simulation are welcome. The type of material considered for publication includes:

1. Research monographs
2. Lectures on a new field or presentations of a new angle in a classical field
3. Summer schools and intensive courses on topics of current research.

Texts which are out of print but still in demand may also be considered if they fall within these categories. The timeliness of a manuscript is sometimes more important than its form, which may be preliminary or tentative.

Titles from this series are indexed by Scopus, Web of Science, Mathematical Reviews, and zbMATH.

Hélène Esnault

Local Systems in Algebraic-Arithmetic Geometry

Hélène Esnault
Institut für Mathematik
Freie Universität Berlin
Berlin, Germany

ISSN 0075-8434 ISSN 1617-9692 (electronic)
Lecture Notes in Mathematics
ISBN 978-3-031-40839-7 ISBN 978-3-031-40840-3 (eBook)
https://doi.org/10.1007/978-3-031-40840-3

This Springer imprint is published by the registered company Springer Nature Switzerland AG
The registered company address is: Gewerbestrasse 11, 6330 Cham, Switzerland

Paper in this product is recyclable.

Contents

Chapter 1
Lecture 1: General Introduction

The *topological fundamental group* $\pi_1(M, m)$ of a connected finite CW-complex M based at a point m, as defined by Poincaré, is a finitely presented group. In turn, any finitely presented group is the fundamental group of a connected finite CW-complex. The finite generation enables one to define a "moduli" (parameter) space $M_B(\pi_1(M, m), r)$ of all its semi-simple complex linear representations $\rho : \pi_1(M, m) \to GL_r(\mathbb{C})$ in a given rank r, modulo conjugation, or equivalently, of all its rank r semi-simple complex local systems $\mathbb{L}$. It is called the character variety of $\pi_1(M, m)$, also the *Betti moduli space* of M in rank r, as conjugation washes out the choice of m, and is a scheme of finite type defined over the ring of integers $\mathbb{Z}$.

We are interested in the case when M consists of the complex points $X(\mathbb{C})$ of a smooth connected algebraic quasi-projective variety X of finite type over the complex numbers $\mathbb{C}$, in which case M has the homotopy type of a connected finite CW-complex. We know extremely little about the restrictions the algebraic origin of $M = X(\mathbb{C})$ imposes on the topological fundamental group $\pi_1(X(\mathbb{C}), x(\mathbb{C}))$. On the other hand, there are naturally defined local systems $\mathbb{L}$, namely those which upon restriction to some Zariski dense open $U \hookrightarrow X$ are subquotients (equivalently, summands by Deligne's semi-simplicity theorem) of a local system on U which comes from the variation of the cohomology of the fibres of a smooth projective morphism $g : Y \to U$. Such $\mathbb{L}$ are called *geometric*. One such example is when $U = X$ and g is in addition *finite* in the topological sense. Then the monodromy of $\mathbb{L}$, that is the group $\mathrm{Im}(\rho)$, defined up to conjugacy, is finite. By the Riemann existence theorem this is equivalent to g being finite étale, and thus relates $\pi_1(X(\mathbb{C}), x(\mathbb{C}))$ to its profinite completion $\pi_1(X_{\mathbb{C}}, x_{\mathbb{C}})$. The latter is the étale fundamental group, defined by Grothendieck, itself related to the Galois group of the field of functions of $X_{\mathbb{C}}$.

So it is natural to try to single out complex points of $M_B(X, r)$ which correspond to geometric or even finite local systems. More generally, it is natural to try to define a notion of geometric sublocus of higher dimension. It is clearly an inaccessible task,

H. Esnault, *Local Systems in Algebraic-Arithmetic Geometry*, Lecture Notes in Mathematics 2337, https://doi.org/10.1007/978-3-031-40840-3_1

which is reminiscent of the Hodge and the Tate conjectures: how can we construct g out of $\mathbb{L}$? There are several *conjectures* relying on various aspects of $M_B(X, r)$.

Grothendieck's p-curvature Conjecture It relies on the *Riemann-Hilbert correspondence* which equates the complex points $\mathbb{L}$ of $M_B(X, r)$ with algebraic semi-simple integrable connections (E, ∇) on X (say X projective for simplicity to avoid boundary growth conditions): we consider (E, ∇) mod p for all large primes p and require this characteristic $p > 0$ connection to be generated by flat sections. This is the original formulation and should characterize finite local systems $\mathbb{L}$. More generally, to characterize geometric local systems $\mathbb{L}$, we request (E, ∇) mod p for all large p to be filtered so that the associated graded is spanned by flat sections. Since the work by Katz which roughly (a bit less) shows that on a geometric $\mathbb{L}$ we can characterize its finiteness by the generation of (E, ∇) mod p by its flat section for almost all p, since the later work by Chudnovsky-Chudnovsky, Bost and André which handle the solvable case, and some remarks of the type made with Kisin, there has been essentially no major progress on this viewpoint.

We discuss in Chap. 2 one possible origin of Grothendieck's p-curvature conjecture, as we understand it, by relating it to the classical rationality criteria of Kronecker. Further in this Chapter, we write a simplified form of Grothendieck's p-curvature conjecture which is equivalent to the general form. We also discuss Katz' proof, giving a slightly different viewpoint.

Gieseker-de Jong Conjecture It relies simply on the *finite generation* of the topological fundamental group $\pi_1(X(\mathbb{C}), x(\mathbb{C}))$, which implies the theorem of Malčev-Grothendieck saying that the étale fundamental group $\pi_1(X_\mathbb{C}, x_\mathbb{C})$ controls the size of $M_B(X, r)$: if $\pi_1(X_\mathbb{C}, x_\mathbb{C}) = \{1\}$ then $M_B(X, r)$ consists of one point, the trivial local system of rank r (in fact there are no extensions as well). Gieseker's conjecture, solved with Mehta, asserts an analog in characteristic $p > 0$ for infinitesimal crystals, while de Jong's conjecture, which is still unsolved in its generality (we understood small steps with Shiho) predicts an analog for isocrystals. It is also related to the Langlands program: if the ground field is $\bar{\mathbb{F}}_p$ and the isocrystal is endowed with a Frobenius structure, then the existence of ℓ-adic companions (as proven with Abe, see also Kedlaya) initially predicted by Deligne in Weil II proves the conjecture. It would be of interest to understand a generalization of de Jong's conjecture on prismatic crystals which encompasses his initial formulation.

We mention in Chap. 3 the proof of the Malčev-Grothendieck theorem. It relies on the finite generation property of the topological fundamental group. We sketch our proof with Mehta of the Gieseker conjecture when X is smooth projective. Although the geometric fundamental group of X is topologically finitely generated, the proof does not use directly this property. Instead it uses the boundedness of Frobenius divided sheaves.

Topological Properties of the (Tame) Fundamental Group of a Smooth (Quasi-) Projective Variety X Defined Over an Algebraically Closed Field k of Characteristic p > 0 As already mentioned, if k was the field of complex numbers, then the topological fundamental group of X would be finitely presented. By Lefschetz theory and Grothendieck's specialization homomorphism from characteristic 0 to characteristic $p > 0$, it is easy to see that the (tame) fundamental group of X is finitely generated as a topological group. But more is true, it is also finitely presented, at least if we assume that X has a good normal crossings compactification. This is a strong analogy with the classical topological situation.

We present in Chap. 4 our proof of it, joint with Shusterman and Srinivas. It rests on Lubotzky's remarkable theorem which characterizes cohomologically this property. The characterization has a motivic flavor in that it says that the H^2 of the (tame) fundamental group with values in rank r continuous representations with $\mathbb{F}_\ell$-coefficients grows linearly in r but does not see ℓ. Indeed, the proof of the independence of ℓ ultimately relies on Deligne's purity from the Weil conjectures. The proof of the linearity in r for $\ell = p$ is more geometric, and, in case X is not proper, uses the existence of a good compactification to have a numerical characterization of tameness.

By the Riemann existence theorem over $\mathbb{C}$ and the base change theorem, the fundamental group of a smooth (quasi-)projective variety over an algebraically closed field of characteristic 0 is the profinite completion of a finitely presented abstract group. We can adapt the notion of finite generation or presentation coming from a discrete group by removing the condition at p. This definition is shaped on the properties of Grothendieck's specialization homomorphism. It yields the concept of p'-finitely generated or presented group. This property, even solely at the level of the p'-finite generation, is an obstruction for X to lift to characteristic 0.

We present in Chap. 5 our proof of this newly defined obstruction, joint with Srinivas and Stix. Ultimately it relies again on a motivic property which this time is not always verified: the representation of the automorphism group of X on its (first) ℓ-adic cohomology is independent of ℓ but is not (always) defined over $\mathbb{Q}$.

Density of Special Loci For X smooth quasi-projective over $\mathbb{C}$, Drinfeld analyzed some arithmetic properties of the Betti moduli space $M_B(X, r)$ viewed as a scheme over $\mathbb{Z}$. At good closed points of $M_B(X, r)$ of characteristic ℓ, de Jong's conjecture applied to the mod p reduction of X for p large predicts the existence of many deformations of the residual representation defined by the closed point to an ℓ-adic sheaf which is arithmetic, that is which is acted on by a power of the Frobenius. Due to the arithmetic Langlands program, those sheaves then are pure in the sense of Deligne, so for example they obey the Hard Lefschetz property if they are semi-simple and X is projective. De Jong's conjecture has been proven by Böckle-Khare in special cases and by Gaitsgory in general (for $\ell \geq 3$) using the geometric Langlands program. It makes it then possible to derive density of certain subloci of $M_B(X, r)$ using those (deep) arithmetic methods.

We present in Chap. 6 a proof, joint with Kerz, that the set of complex points of $M_B(X, r)$ corresponding to semi-simple local systems with quasi-unipotent monodromy at infinity is Zariski dense. It uses Drinfeld's idea. This is an invitation to transpose the definition of arithmetic ℓ-adic local systems on the mod p reduction of X to a notion of weakly arithmetic complex local systems on X over $\mathbb{C}$. We present the definition and the proof, joint with de Jong, that the set of weakly arithmetic local systems in $M_B(X, r)$ is Zariski dense. All weakly arithmetic local systems have quasi-unipotent monodromies at infinity, so this density is sharper than the previous one. The way to go from a complex local system to an ℓ-adic étale local system in the definition prevents us to conclude that there only countably many weakly arithmetic local systems.

On the other hand, Biswas-Gupta-Mj-Whang in rank 2, resp. Landesman-Litt in any rank proved by topological, resp. Hodge theoretical methods that on a geometric generic curve of genus ≥ 2, semi-simple complex local systems of low rank which descend to the universal curve are unitary.

Granted this, we present in Chap. 6 the idea of their proof in any rank to the effect that in fact not only those local systems are unitary, but they have finite monodromy. The proof ultimately relies on the integrality of cohomologically rigid local systems, a theorem joint with Groechenig, explained in Chap. 7. It also shows that local systems on the geometric generic curve which come from the universal curve cannot be dense in the Betti moduli, contrary to a hope I had expressed with Kerz earlier on.

Companions and Integrality A complex local system can be twisted by an automorphism σ of the field of complex numbers: we post-compose the underlying linear representation of the topological local system by σ on the coefficients. If we now consider representations of a topological group with values in a linear group with coefficients in a topological field, we lose continuity by post-composing. So we cannot define twisted continuous representations. Deligne in Weil II predicted that nonetheless it is possible in the following situation. We assume that the topological group is the geometric fundamental group of a smooth quasi-projective variety X defined over a finite field $\mathbb{F}_q$, the field of coefficients is the algebraic closure $\bar{\mathbb{Q}}_\ell$ of the ℓ-adic numbers for some prime number ℓ prime to p, the isomorphism σ is some abstract isomorphism between $\bar{\mathbb{Q}}_\ell$ and $\bar{\mathbb{Q}}_{\ell'}$ for some prime number ℓ' prime to p. If the local system is stabilized by some power of the Frobenius, that is if it is arithmetic, then by the Langlands correspondence the characteristic polynomial of the action of the Frobenius elements at closed points is a polynomial with coefficients being algebraic numbers. Thus σ sends them to other algebraic numbers. In addition, a simple ℓ-adic local system is recognized by the Čebotarev density theorem by those polynomials. Deligne then conjectured the existence of a simple ℓ'-adic local system with those data, which he called (σ)-companion. If we believe that arithmetic local systems come from geometry, then we know at least on the whole Gauß-Manin system how to go from ℓ to ℓ'. If X has dimension 1, as a corollary of the Langlands correspondence, L. Lafforgue proved the existence of companions the way Drinfeld did after he proved the rank 2 case of the Langlands program. In addition, arithmetic local systems come from geometry. There is no Langlands program in higher dimension. In absence of any geometric support, Drinfeld proved the existence of companions in higher dimension as well.

We present in Chap. 7 some aspects of Drinfeld's proof. The starting point is the dimension 1 case due to L. Lafforgue, see above. The problem is how to glue those ℓ'-adic sheaves defined on all curves, which agree on intersections. Drinfeld produces a non-commutative version of the construction performed by Wiesend. He uses a strong form of the Čebotarev density theorem: an ℓ'-adic local system on a curve is recognized on finitely many points. He also uses Deligne's earlier work showing that for a given ℓ-adic local system, the coefficients of the characteristic polynomials of the Frobenii at closed points stay in a finite type extension of $\mathbb{Q}$.

Simpson's Motivicity Conjecture: Rigid Local Systems These are the 0-dimensional components of $M_B(X, r)$. Simpson predicted that they are all geometric. It relies on the corresponding theorem by Katz when X has dimension 1, in which case X has to be an open in $\mathbb{P}^1$ (so in the definition of $M_B(X, r)$ one has to fix conjugacy classes of quasi-unipotent monodromy at infinity). It relies also on the Simpson correspondence, which, when X is projective, equates real analytically $M_B(X, r)$ with the moduli space of semi-stable Higgs bundles with vanishing Chern classes. Those are endowed with a $\mathbb{C}^\times$-flow, thus rigid local systems, viewed on the Higgs side, are fixed by it. This implies, according to Simpson's theorem, that they underly a polarizable complex variation of Hodge structures. From there it is one short step to dream of geometricity.

We present in Chap. 7 our proof with Groechenig of a consequence of the motivicity conjecture, called the integrality conjecture, also formulated by Simpson: rigid local systems should be integral. To this aim we descend the ℓ-adic local system mod p for p very large. The rigidity implies that this ℓ-adic local system is in fact arithmetic as well, a fact already observed by Simpson. We then take a companion, and interpret back this ℓ'-local system topologically on the initial complex variety. To be able to conclude we need that this newly defined topological system is rigid as well. We can prove this only under the extra assumption that the rigid local system we started with was a smooth point in its moduli. This is a cohomological condition which we call cohomological rigidity.

Integrality of the Betti Moduli Space An important point is that rigid local systems are arithmetic. On the other hand, as mentioned above as a consequence of de Jong's conjecture, weakly arithmetic local systems are dense.

We present in Chap. 7 our proof with de Jong of an integrality property of the whole open of the Betti moduli corresponding to irreducible local systems. The proof rests on this notion of weakly arithmetic local systems together with the same idea to prove integrality of (cohomologically) rigid local systems. One consequence of the property is that it yields a new obstruction, visible on the Betti moduli, for a finitely presented group to be the topological fundamental group of a smooth complex quasi-projective variety.

Crystallinity Property of Rigid Local Systems If we believe in Simpson's motivicity conjecture, it should have two p-adic consequences, one on the local system $\mathbb{L}$ itself, the other one on the flat connection (E, ∇) which is defined using the Riemann-

Hilbert correspondence. Indeed if those data ultimately come from a splitting of a Gauß-Manin system of a smooth projective morphism $g : Y \to U$ on some dense open $U \hookrightarrow X$, then for p large all the data have good reduction. So one expects that the p-completion of the local system on a p-adic model of X with good reduction, for p large, is crystalline. At the same time, if g does not have good reduction but X has, one expects the connection, in restriction to the p-adic model, to yield an isocrystal with a Frobenius structure. Those two properties, with less precision on the "good" p, were proven with Groechenig.

In Chap. 8 we present a proof of the F-isocrystal property which relies on a theorem of Berthelot which is nearly documented in the Stack Project and which I learned from de Jong: if the integral p-adic model of X is unramified, and $p \geq 3$, then the Frobenius acts on connections defined on it. This enables one to prove a posteriori that the p-curvature of the mod p reduction of the connection is nilpotent. Our original proof with Groechenig showed this property first, which was a bit more difficult. For the crystallinity of the p-adic local system however, in the present state of knowledge, we have to return to our proof with Groechenig which proceeds along the completion at p by showing that we have a periodic Higgs-de Rham flow in the sense of Zuo and his coauthors.

It is to be noted that this crystallinity proof is easily generalized to non-proper smooth varieties under a strong cohomological condition. We do not reproduce our proof with Groechenig in those notes. We observe that precisely this property is the one used by Pila-Shankar-Tsimerman to prove the André-Oort conjecture on Shimura varieties of real rank ≥ 2.

There are manifold topics close to the ones discussed in those notes which are not addressed. One of them concerns specific subloci of the Betti moduli space which could be called arithmetic and which we defined with Kerz. It generalizes to higher dimensional subvarieties of the Betti moduli the notion of arithmeticity of an integral ℓ-adic point. In rank 1 we proved with Kerz that those are translates by torsion points of subtori in the whole moduli which itself is a torus. We could not progress in higher rank in this direction, for lack of understanding the more precise geometry of the moduli. Another direction is what to expect after the work of Petrov who showed that the generalized Fontaine-Mazur conjecture could simply be formulated by predicting that geometrically irreducible ℓ-adic local systems with trivial determinant which are arithmetic on X smooth quasi-projective in characteristic 0 should in fact come from geometry. It would be of help to find one small consequence of this vast conjecture, as a reality check.

Finally in Chap. 10 we formulate a few questions and problems related to the various Lectures.

Acknowledgements

Special heartfelt thanks go to Michael Groechenig, Moritz Kerz and more recently Johan de Jong with whom I developed part of the mathematics presented here. Those notes hugely reflect ideas we developed together and further share.

The influence of the ideas of Pierre Deligne and Vladimir Drinfeld is overwhelming and permeates the whole edifice of ideas exposed in those notes. I thank Pierre Deligne for a thorough list of comments and remarks on a first draft of those notes.

I thank Tomoyuki Abe, Lars Kindler, Mark Kisin, Adrian Langer, Vikram Mehta, who unfortunately is no longer among us, Atsushi Shiho, Mark Shusterman, Vasudevan Srinivas, Jakob Stix for the joint work reproduced in those notes.

More specifically, I thank Piotr Achinger, Yves André, Benjamin Bakker, Alexander Beilinson, Barghav Bhatt, Emmanuel Breuillard, Yohan Brunebarbe, Benjamin Church, Dustin Clausen, Marco D'Addezio, Pierre Deligne, Johan de Jong, Vladimir Drinfeld, Mikolaj Fraczyk, Michael Groechenig, Michael Harris, Lars Hesselholt, Ben Heuer, Moritz Kerz, Mark Kisin, Bruno Klingler, Raju Krishnamoorty, Aaron Landesman, Adrian Langer, Daniel Litt, Alexander Lubotzky, Akhil Mathew, John Morgan, Alexander Petrov, Jonathan Pila, Maxime Ramzi, Vasily Rogov, Will Sawin, Ananth Shankar, Carlos Simpson, Jacob Tsimerman, Vadim Vologodsky, Kang Zuo. I had with them exchanges at various stages, they all contributed in this way to the elaboration of the Lecture Notes.

I thank the various referees for the work they accomplished. It is a privilege to have the Lecture Notes be read with attention. I took care of all the suggestions. I was also touched by the warmth of the general comments.

I thank the mathematicians from Columbia University. They kindly invited me to give the Eilenberg Lectures in the fall 2020 which are the support of those notes. Because of the pandemic, I "used up" three chairs, Michael Thaddeus, Robert Friedman and Johan de Jong. I am sorry for the work it created for them. Finally I came during Johan's "reign" in the fall 2022. It has been a wonderful time.

I thank the mathematicians of the University of Copenhagen. I gave there an echo of the Eilenberg Lectures in the fall 2021 and the winter 2023.

I thank all the students, post-docs and Faculty who listened to the lectures and contributed to make them colorful and I hope enjoyable. I also thank the administrative staff of both institutions for warmly welcoming me.

I thank Springer Verlag which allowed me to write the notes in a colloquial way, Lecture 1, Lecture 2 etc. To understand the details of some proofs, we always have to go back to the original literature. This is all the more true for those notes. The point to write them is to convey a philosophy and a program which I developed partly with other mathematicians within the last 10 or 15 years, and to share some dreams. Sharing dreams can only be done in an intimate style.

Acknowledgments

[illegible]

Chapter 2
Lecture 2: Kronecker's Rationality Criteria and Grothendieck's *p*-Curvature Conjecture

Abstract We recall two criteria, say an analytic one and an algebraic one, for an integral number to be a root of unity and for an algebraic number to be a rational number. Both go back to Kronecker. We recall Grothendieck's p-curvature conjecture and its generalization, avoiding the general definition of the p-curvature of a connection in characteristic $p > 0$. We show how the two are related and mention Katz's proof using (a generalization of) the analytic criterion.

2.1 Kronecker's Criteria

Let $a \in \mathbb{C}$ be a complex number, and write it as $a = \exp(2\pi\sqrt{-1}b)$ for $b \in \mathbb{C}$ defined modulo the integers $\mathbb{Z}$. We list Kronecker's criteria for $a \in \mu_\infty \subset \mathbb{C}$, that is for a to be a root of unity, or equivalently for $b \in \mathbb{Q}$, that is for b to be a rational number.

2.1.1 Kronecker's Analytic Criterion, [Kro57]

Recall that the subring $\bar{\mathbb{Z}} \subset \mathbb{C}$ of algebraic integers of the complex numbers consists of those complex numbers $a \in \mathbb{C}$ satisfying an equation $f(a) = a^d + c_1a^{d-1} + \ldots + c_d = 0$ with $c_i \in \mathbb{Z}$, and the subfield $\bar{\mathbb{Q}} \subset \mathbb{C}$ of algebraic numbers consists of those a as before but with $c_i \in \mathbb{Q}$. Any field automorphism $\sigma \in \mathrm{Aut}(\mathbb{C})$ of $\mathbb{C}$ leaves $\bar{\mathbb{Z}} \subset \bar{\mathbb{Q}}$ invariant.

Proposition 2.1 *Assuming $a \in \bar{\mathbb{Z}}$, then $a \in \mu_\infty$ if and only if for any $\sigma \in \mathrm{Aut}(\mathbb{C})$, the complex absolute value of $\sigma(a)$ is equal to* 1.

Proof See https://mathoverflow.net/questions/10911/english-reference-for-a-result-of-kronecker. For $f = f_1$ as above, we write $f_n(X) = \prod_{i=1}^{d}(X - \alpha_i^n) \in \mathbb{C}[X]$ for all $n \in \mathbb{N}_{>0}$, with $\alpha_1 = a$. The coefficients of f_n are symmetric functions in the α_i, thus are expressable as polynomials with rational coefficients in the c_i, and on the

H. Esnault, *Local Systems in Algebraic-Arithmetic Geometry*, Lecture Notes in Mathematics 2337, https://doi.org/10.1007/978-3-031-40840-3_2

other hand they are in $\bar{\mathbb{Z}}$, thus they lie in $\mathbb{Z}$, and have bounded norms. Thus there are finitely many such f_n, thus the set $\{a^n, n \in \mathbb{N}_{\geq 1}\}$ is finite, thus lies in μ_∞. □

2.1.2 *Kronecker's Algebraic Criterion, [Bau04]*

Assume $b \in \bar{\mathbb{Q}}$, so b lies in the number field $\mathbb{Q}(b)$, of rank d say. So we can take its reduction modulo almost all primes, that is modulo all except finitely many primes $\mathfrak{p}$ of the number ring $\mathcal{O}_{\mathbb{Q}(b)}$.

Proposition 2.2 *If for almost all primes $\mathfrak{p}$ of $\mathcal{O}_{\mathbb{Q}(b)}$, $(b \bmod \mathfrak{p}) \in \mathbb{F}_p \subset \mathcal{O}_{\mathbb{Q}(b)}/\mathfrak{p}$, then $b \in \mathbb{Q}$. In words: if b lies in the prime field in characteristic p for almost all p, then it does in characteristic* 0.

Proof After multiplying b by an element in $\mathbb{Z} \setminus \{0\}$, we may assume that $b \in \mathcal{O}_{\mathbb{Q}(b)}$, so generates the subring $\mathbb{Z}[b] \subset \mathcal{O}_{\mathbb{Q}(b)}$ which is free of rank d over $\mathbb{Z}$, which is also the rank of $\mathcal{O}_{\mathbb{Q}(b)}$ over $\mathbb{Z}$. Hence for all but finitely many primes $\mathfrak{p}$ of $\mathcal{O}_{\mathbb{Q}(b)}$, $\mathcal{O}_{\mathbb{Q}(b)}/\mathfrak{p}$ is spanned by $(b \bmod \mathfrak{p})$ over $\mathbb{F}_p$. So the condition $(b \bmod \mathfrak{p}) \in \mathbb{F}_p \subset \mathcal{O}_{\mathbb{Q}(b)}/\mathfrak{p}$ is equivalent to $(b \bmod \mathfrak{p})$ being completely split. By Kronecker's density theorem, this implies $d = 1$.

□

2.1.3 *Translation of Kronecker's Algebraic Criterion in Terms of Differential Equations*

We consider the complex algebraic variety

$$X = \operatorname{Spec} \mathcal{O}(X), \ \mathcal{O}(X) = \mathbb{C}[t, t^{-1}]$$

and on it the linear differential equation

$$(\star) \quad \frac{df}{f} = b\frac{dt}{t}$$

for some $b \in \mathbb{C}$. It has the analytic solution

$$f_\lambda(t) = \lambda t^b, \ \lambda \in \mathbb{C}.$$

Then

Lemma 2.3 *For $\lambda \neq 0$, f_λ is algebraic over $\mathcal{O}(X)$ if and only if $b \in \mathbb{Q}$ if and only if f_λ is integral over $\mathcal{O}(X)$.*

Proof If $\mathbb{Q} \ni b = \frac{m}{n}$, $m, 0 \neq n \in \mathbb{Z}$ then $f_\lambda^n \in \mathcal{O}(X)$ so f_λ is integral over $\mathcal{O}(X)$. Assume now f_λ is algebraic over $\mathcal{O}(X)$, so in particular f_λ is algebraic over the Laurent power series field $\mathbb{C}((t))$ containing $\mathcal{O}(X)$. So f_λ defines a finite field extension

$$\mathbb{C}((t)) \hookrightarrow \mathbb{C}((t))(f_\lambda).$$

As the embedding

$$\cup_{n\in\mathbb{N}} \mathbb{C}((t^{1/n})) \hookrightarrow \overline{\mathbb{C}((t))}$$

is an equality, $C((t))(f_\lambda)$ must be one of the fields $\mathbb{C}((t^{1/n}))$ thus $b \in \mathbb{Q}$, which also implies that f_λ is integral over $\mathcal{O}(X)$.

□

Remark 2.4 For $\lambda \neq 0$, the following conditions are equivalent:

(i) f_λ is algebraic over the field of fractions $\text{Frac}(\mathcal{O}(X)) = \mathbb{C}(t)$;
(ii) f_λ is integral over $\mathcal{O}(X)$;
(iii) the monodromy of $(\star)$ is finite.

Proof (i) $\Longrightarrow$ (ii): f_λ, as a solution of $(\star)$ is analytic on X, and by the condition (i), f_λ lies in $\overline{\mathbb{C}(t)}$. Thus f_λ lies in the integral closure $\overline{\mathcal{O}(X)} \subset \overline{\mathbb{C}(t)}$.

(ii) $\Longrightarrow$ (iii): Recall that by Lemma 2.3, $f_\lambda = \lambda t^b$. As f_λ is integral over $\mathcal{O}(X)$, it is integral over $\mathbb{C}((t))$. So by Kummer theory, $b \in \mathbb{Q}$. On the other hand, the restriction $\gamma^* f_\lambda$ of f_λ to the path $\big(\gamma : [0\ 1] \to, \tau \mapsto \exp(2\pi\sqrt{-1}\tau)\big)$ is the function $\big([0\ 1] \to \mathbb{C},\ \tau \mapsto \lambda\exp(2\pi\sqrt{-1}b\tau)\big)$. So the monodromy $a \in \mathbb{C}^\times$ on γ is computed as the ratio

$$a = \gamma^* f_\lambda(\tau = 1)/\gamma^* f_\lambda(\tau = 0) = \exp(2\pi\sqrt{-1}b),$$

which is a root of unity.

(iii) $\Longrightarrow$ (i): If a is a root of unity, then b is a rational number so (i) holds.

□

We now assume $b \in \bar{\mathbb{Q}}$. For almost all primes $\mathfrak{p}$ of $\mathbb{Q}(b)$ where it makes sense, that is for which b is integral in the corresponding p-adic field, we consider the differential equation

$$(\star)_\mathfrak{p} \quad \frac{df}{f} = (b \bmod \mathfrak{p})\frac{dt}{t}$$

which is simply $(\star)$ by viewed over $\mathcal{O}_{\mathbb{Q}(b)}/\mathfrak{p}$. We write $\mathcal{O}_{\mathbb{Z}}(X) = \mathbb{Z}[t, t^{-1}]$ so $\mathcal{O}(X) = \mathcal{O}_{\mathbb{Z}} \otimes_{\mathbb{Z}} \mathbb{C}$. The only way to make sense of the solutions

$$\lambda t^b \text{ for } \lambda \in \mathcal{O}_{\mathbb{Q}(b)}/\mathfrak{p}$$

is to request them to lie in $\mathcal{O}_{\mathbb{Z}}(X) \otimes_{\mathbb{Z}} (\mathcal{O}_{\mathbb{Q}(b)}/\mathfrak{p})$. For $\lambda \neq 0$ in $\mathcal{O}_{\mathbb{Q}(b)}/\mathfrak{p}$, λt^b lies in $\mathcal{O}_{\mathbb{Z}}(X) \otimes_{\mathbb{Z}} (\mathcal{O}_{\mathbb{Q}(b)}/\mathfrak{p})$ if and only if (b mod $\mathfrak{p}$) lies in $\mathbb{F}_p \subset \mathcal{O}_{\mathbb{Q}(b)}/\mathfrak{p}$.

We now summarize the discussion.

Theorem 2.5 *If $b \in \bar{\mathbb{Q}}$, the differential equation $(\star)$ has one non-trivial algebraic solution over $\mathcal{O}(X)$ if and only if for almost all primes $\mathfrak{p} \in \mathcal{O}_{\mathbb{Q}(b)}$, $(\star)_{\mathfrak{p}}$ has one non-trivial solution in $\mathcal{O}_{\mathbb{Z}}(X) \otimes_{\mathbb{Z}} (\mathcal{O}_{\mathbb{Q}(b)}/\mathfrak{p})$.*

Remark 2.6 The last condition, to have one non-trivial solution, or equivalently, as the differential equation has rank 1, to have a full set of solutions in $\mathcal{O}_{\mathbb{Z}}(X) \otimes_{\mathbb{Z}} (\mathcal{O}_{\mathbb{Q}(b)}/\mathfrak{p})$, by definition is equivalent to saying that the p-curvature of the differential equation $(\star)_{\mathfrak{p}}$ vanishes. We do not give here the definition of the p-curvature itself, as we do not need it as such, we only need what it means for it to vanish (or later in the Lecture to be nilpotent).

Proof of Theorem 2.5 $(\star)$ has a full set of algebraic solutions over $\mathcal{O}(X)$

if and only if

$b \in \mathbb{Q}$ (Lemma 2.3)

if and only

(b mod $\mathfrak{p}$) $\in \mathbb{F}_p \subset (\mathcal{O}_{\mathbb{Q}(b)}/\mathfrak{p})$ for almost all p (Proposition 2.2)

if and only

$(\star)_{\mathfrak{p}}$ has a full set of solutions in $\mathcal{O}_{\mathbb{Z}}(X) \otimes_{\mathbb{Z}} (\mathcal{O}_{\mathbb{Q}(b)}/\mathfrak{p})$ (previous discussion). □

The content of Grothendieck's p-curvature conjecture is to predict that Theorem 2.5 extends to any smooth complex quasi-projective variety X and any linear differential equation $(\star)$. We shall explain the formulation and a generalization of it without ever mentioning the definition of the p-curvature.

2.2 Grothendieck's p-Curvature Conjecture

Let us first formulate what *a posteriori* is equivalent to the p-curvature conjecture.

We write $X = \text{Spec}\mathcal{O}(X)$, $X = \text{Spec}\mathbb{C}[t, t^{-1}, (t-1)^{-1}]$. Then we pose the linear differential equation

$$(\star) \quad \frac{df}{f} = b\frac{dt}{t} + c\frac{dt}{t-1}$$

where now

$$b = (b_{ij}),\ c = (c_{ij}) \in M_r(\bar{\mathbb{Q}}).$$

For almost all primes $\mathfrak{p}$ of $\mathbb{Q}(b_{ij}, c_{ij}),\ 1 \leq i, j \leq r$ where it makes sense, we consider the differential equation

$$(\star)_{\mathfrak{p}} \quad \frac{df}{f} = (b \bmod \mathfrak{p})\frac{dt}{t} + (c \bmod \mathfrak{p})\frac{dt}{t-1}$$

which is simply $(\star)$ but viewed over $\mathcal{O}_{\mathbb{Q}(b_{ij})}/\mathfrak{p}$. We write $\mathcal{O}_{\mathbb{Z}}(X) = \mathbb{Z}[t, t^{-1}, (t-1)^{-1}]$ so $\mathcal{O}(X) = \mathcal{O}_{\mathbb{Z}}(X) \otimes_{\mathbb{Z}} \mathbb{C}$.

We say that $(\star)_p$ has a full set of solutions if it has a set of r solutions

$$(f_1, \dots, f_r),\ f_i \in \mathcal{O}_{\mathbb{Z}}(X) \otimes_{\mathbb{Z}} (\mathcal{O}_{\mathbb{Q}(b_{ij})}/\mathfrak{p}),$$

which are linearly independent over $\mathcal{O}_{\mathbb{Q}(b_{ij})}/\mathfrak{p}$. Similarly we say that $(\star)$ has a full set of algebraic solutions if it has a set of r solutions $(f_1, \dots, f_r)$ where f_i is algebraic over $\mathcal{O}(X)$, and which are linearly independent over $\mathbb{C}$.

Conjecture 2.7 The differential equation $(\star)$ has a full set of algebraic solutions over $\mathcal{O}(X)$ if and only if $(\star)_{\mathfrak{p}}$ has a full set of solutions in $\mathcal{O}_{\mathbb{Z}}(X) \otimes_{\mathbb{Z}} (\mathcal{O}_{\mathbb{Q}(b_{ij})}/\mathfrak{p})$ (that is its p-curvature vanishes).

The initial Grothendieck's conjecture is documented in Katz's article [Kat72, Introduction]: let X be a smooth quasi-projective variety over $\mathbb{C}$, (E, ∇) be an integrable algebraic connection. Let S be a scheme of finite type over $\mathbb{Z}$ so $(X, (E, \nabla))$ descends to $(X_S, (E_S, \nabla_S))$ and has good reduction at all closed points $s \in S$. Then the prediction is the following conjecture.

Conjecture 2.8 (Grothendieck's p-Curvature Conjecture) (E, ∇) has a full set of algebraic solutions if and only there is a dense open $S^\circ \subset S$ such that for all $s \in S^\circ$ the reduction (E_s, ∇_s) on X_s has a full set of solutions (that is its p-curvature vanishes).

Aside of the viewpoint presented here, another origin of the conjecture is [Lan04].

We explain briefly why Conjecture 2.8 is equivalent to its special case Conjecture 2.7.

Claim 2.9 ([Kat70], Theorem 13.0) If there is a dense open $S^\circ \subset S$ such that for all $s \in S^\circ$ the reduction (E_s, ∇_s) has a full set of solutions, then (E, ∇) is regular singular and has finite monodromies at infinity.

Claim 2.10 (E, ∇) has a full set of algebraic solutions if and only if there is a finite cover $h' : Y' \to X$ such that $h'^*(E, \nabla)$ has a full set of solutions, if and only if there is a finite unramified cover $h : Y \to X$ such that $h^*(E, \nabla)$ has a full set of solutions.

Proof If h exists, the solutions of (E, ∇) in $\mathcal{O}(Y)$ are algebraic solutions over $\mathcal{O}(X)$. Vice-versa, assume given a basis over $\mathcal{O}(X)$ of algebraic solutions f_i say. The ring $\mathcal{O}(X) \hookrightarrow \mathcal{O}(X)[f_i]$ is then finite, thus defines a finite possibly ramified cover $h' : Y' \to X$ so $h'^*(E, \nabla)$ is trivial. So the underlying monodromy representation $\pi_1(X) \to GL_r(\mathbb{C})$ of (E, ∇) trivializes on $h'_*\pi_1(Y')$ in $\pi_1(X)$, which has finite index. This defines a finite unramified cover $h : Y \to X$ with $\pi_1(Y) = h'_*\pi_1(Y')$ such that $h^*(E, \nabla)$ is trivial.

□

Claim 2.11 Conjecture 2.8 in general is equivalent to its special case when X is a smooth projective curve.

Proof By Claim 2.10, Conjecture 2.8 is true for (E, ∇) on X if and only if it is for $h'^*(E, \nabla)$ for any $h' : Y' \to X$ finite cover. So by Claim 2.9 we may assume that (E, ∇) extends to a smooth projective compactification of X. Then the Lefschetz hyperplane theorem reduces it to the case of a smooth projective curve.

□

Claim 2.12 ([And04], Théorème 0.6.1) Conjecture 2.8 is equivalent to its special case when X is a smooth projective curve defined over a number field.

In other words, let $f : X \to S$ be a smooth morphism between smooth quasi-projective varieties defined over a number field, let (E, ∇) be a relative flat connection on X. If for all closed points $s \in S$, the monodromy of the restriction $(E, \nabla)|_{X_s}$ of (E, ∇) to the fibre X_s is finite, that is if there is a finite étale cover $h_s : Y_s \to X_s$ such that $h_s^*(E, \nabla)|_{X_s}$ is trivial, then the monodromy of (E, ∇) restricted to a geometric generic fibre $X_{\bar{\eta}}$ is finite as well. We remark that the arguments can be adapted to prove a characteristic $p > 0$ version of the same theorem ([EL13, Theorem 5.1]).

Applying Belyi's theorem [Bel80], we formulate the following claim.

Claim 2.13 Conjecture 2.8 is equivalent to its special case when X is $\mathbb{P}^1 \setminus \{0, 1, \infty\}$, that is Conjecture 2.7. defined over a number field.

2.3 Back to Kronecker's Analytic Criterion: Gauß-Manin Connections

Let X be a quasi-projective smooth variety over $\mathbb{C}$, $g : Y \to X$ be a smooth projective morphism, $\mathbb{L}$ being the local system $R^i g_* \mathbb{C}$ for some i. Then

Theorem 2.14 ([Kat72]) *Conjecture 2.8 holds for* $\mathbb{L}$.

Proof The proof consists of two parts. First, the assumption on the full set of solutions for $(\star)_p$ implies that the F-filtration $R^i g_* \Omega^{\geq j}_{Y/X} \hookrightarrow R^i g_* \Omega^{\bullet}_{Y/X}$ is stabilized by the Gauß-Manin connection. Then this implies that the monodromy is an automorphism of the integral polarized Hodge structure of the fibre, which is the intersection of the integral group $\mathrm{Aut}(H^i(Y_x, \mathbb{Z}))$ with $\mathrm{Aut}(H^i(Y_x, \mathbb{R}), (2\pi\sqrt{-1}^i(x, Cy))$ inside of $\mathrm{Aut}(H^i(Y_x, \mathbb{R}))$, where C is the Weil operator. A variant of Kronecker's analytic criterion says that this intersection is finite.

So the stabilization of the F-filtration by the connection is the main point. We give a presentation which slightly differs from *loc. cit.* To this, over our model X_S, we take a closed point s of large characteristic p so the coherent sheaves $R^{i-j}g_*\Omega^j_{Y/X}$ remain locally free, and the map

$$(*) \quad R^i g_* \Omega^{\geq j}_{Y/X} \to R^i g_* \Omega^{\bullet}_{Y/X},$$

which by Hodge theory is injective, remains injective by base change. We want to prove that the condition on the existence of a full set of solutions implies that the F-filtration is stabilized by the connection, that is that the $\mathcal{O}_X$-linear Kodaira-Spencer class

$$KS: R^{i-j}g_*\Omega^j_{Y/X} \xrightarrow{\nabla} \Omega^1_X \otimes R^{i+1-j}g_*(\Omega^{j-1}_{Y/X} \to \Omega^j_{Y/X}) \to \Omega^1_X \otimes R^{i+1-j}g_*\Omega^{j-1}_{Y/X}$$

dies. The important diagram is then

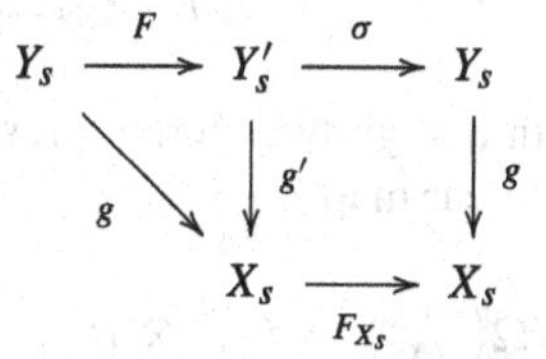

$$\begin{array}{ccccc} Y_s & \xrightarrow{F} & Y'_s & \xrightarrow{\sigma} & Y_s \\ & \searrow^{g} & \downarrow{\scriptstyle g'} & & \downarrow{\scriptstyle g} \\ & & X_s & \xrightarrow[F_{X_s}]{} & X_s \end{array}$$

where $F : Y_s \to Y'_s$ is the Frobenius on Y_s relative to X_s, that is on local functions $\mathcal{O}(Y_s) = \mathcal{O}(X_s)[y_1, \ldots, y_n]/I$ it sends the class of $f(y_1, \ldots, y_n)$ to the class of $f(y_1^p, \ldots, y_n^p)$, and $F_{X_s} : X_s \to X_s$ is the absolute Frobenius, so it sends $f \in \mathcal{O}(X_s)$ to $f^p \in \mathcal{O}(X_s)$. The injectivity of $(*)$ implies that

$$(**) \quad R^i g_* \tau_{\leq j} \Omega^{\bullet}_{Y_s/X_s} = R^i g'_* \tau_{\leq j} F_* \Omega^{\bullet}_{Y_s/X_s} \to R^i g'_* F_* \Omega^{\bullet}_{Y_s/X_s}$$

is injective as well. By the Cartier isomorphism and base change for F_{X_s}, it holds

$$R^{i-j}g_*\mathcal{H}^j = R^{i-j}g'_*\Omega^j_{Y'_s/X_s} = F^*_{X_s} R^{i-j}g_*\Omega^j_{Y_s/X_s},$$

where $\mathcal{H}^j$ is the j-th cohomology sheaf of $F_*\Omega^{\bullet}_{Y_s/X_s}$. Computing the dimension of de Rham cohomology $H^i_{dR}(Y_x)$ of the fibres of g yields that the following sequence of connections

$$(\epsilon) \quad 0 \to F^*_{X_s} R^{i-j}g_*\Omega^j_{Y_s/X_s} \to R^i g_*(\tau_{\leq j+1} F_* \Omega^{\bullet}_{Y_s/X_s} / \tau_{\leq j-1} F_* \Omega^{\bullet}_{Y_s/X_s})$$
$$\to F^*_{X_s} R^{i-j-1} g_* \Omega^{j+1}_{Y_s/X_s} \to 0$$

is exact. It defines a de Rham extension

$$\epsilon \in H^1_{dR}(X_s, F^*_{X_s} R^{i-j-1} g_*(\Omega^{j+1}_{Y_s/X_s})^\vee \otimes F^*_{X_s} R^{i-j} g_* \Omega^j_{Y_s/X_s}).$$

Now we use the assumption which is equivalent to saying that the exact sequence (ϵ) of connections is equal to $F^*_{X_s}$ of an exact sequence

$$0 \to R^{i-j} g_* \Omega^j_{Y_s/X_s} \to V \to R^{i-j-1} g_* \Omega^{j+1}_{Y_s/X_s} \to 0$$

of vector bundles on X_s, endowed with the canonical connection. In particular this sequence splits on a dense affine open of X_s. Thus ϵ dies as well on a dense affine open. On the other hand, the cohomology

$$H^1_{dR}(X_s, F^*_{X_s} R^{i-j-1} g_*(\Omega^{j+1}_{Y_s/X_s})^\vee \otimes F^*_{X_s} R^{i-j} g_* \Omega^j_{Y_s/X_s}) =$$
$$H^1_{dR}(X_s, (R^{i-j-1} g_*(\Omega^{j+1}_{Y_s/X_s})^\vee \otimes R^{i-j} g_* \Omega^j_{Y_s/X_s}) \otimes F_{X_s *}(\Omega^\bullet_{X_s}))$$

maps, again via the Cartier isomorphism, to

$$H^0(X_s, (R^{i-j-1} g_*(\Omega^{j+1}_{Y_s/X_s})^\vee \otimes R^{i-j} g_* \Omega^j_{Y_s/X_s}) \otimes \Omega^1_{X_s}).$$

Katz [Kat72, Theorem 3.2] (in a slightly different presentation) computes that the image of ϵ, viewed as a $\mathcal{O}_{X_s}$-linear map

$$R^{i-j-1} g_*(\Omega^{j+1}_{Y_s/X_s}) \to \Omega^1_{X_s} \otimes R^{i-j} g_* \Omega^j_{Y_s/X_s}$$

is precisely the Kodaira-Spencer map KS_s for g. Thus KS_s dies on a dense open of X_s. As KS_s is a $\mathcal{O}_{X_s}$-linear morphism between vector bundles, we conclude that it dies on the whole of X_s. Thus KS dies in restriction to all closed points in a dense open in S. Thus it dies. This finishes the proof.

□

The works by Chudnovsky-Chudnovsky [Chu85], Bost [Bos01] and André [And04] handle the solvable case. There are also some remarks like [EK18] under a stronger condition than just generation of the differential equation on characteristic p. Else, there is a different idea in Ananth Shankar's work [Sha18] who showed finiteness of the monodromy on simple loops, which enabled him and Patel-Whang to finish the proof in rank 2 in [PSW21]. Beyond this, there is essentially no big progress on this viewpoint.

However, there are two more points. The proof above using the stabilization of the τ-filtration (also called *conjugate filtration*) suggests the generalization:

Conjecture 2.15 (Generalized Grothendieck's p-Curvature Conjecture, [And89], Appendix to Ch. V) To characterize geometric local systems $\mathbb{L}$, we request that the underlying vector bundle of a connection (E, ∇) has the property that mod p

for all large p it is filtered so that the associated graded is spanned by flat sections (that is its p-curvature is nilpotent) on a dense open of X mod p.

Said differently, Katz proves: if mod p for all large p, (E, ∇) is filtered so that the associated graded is spanned by flat sections, and in addition (E, ∇) comes from geometry, and furthermore this filtration is trivial, then (E, ∇) has finite monodromy.

There is one other instance of a similar situation. Let X be a smooth projective variety defined over $\mathbb{C}$ and let us assume that $H^0(X, \mathrm{Sym}^{>0}\Omega^1_X) = 0$. I had conjectured that then all local systems have finite monodromy. This has been proved in [BKT13] by *analytic methods* as a corollary of Zuo's theorem [Zuo00] to the effect that if the derivative of the period map associated to an integral variation of Hodge structure is injective, then Ω^1_X is big. In [BKT13] one finds a large list of examples of such varieties.

Algebraically, the vanishing $H^0(X, \mathrm{Sym}^{>0}\Omega^1_X) = 0$ implies that the underlying bundles with connection (E, ∇) have the property that mod p for all large p they are filtered so that the associated graded is spanned by flat sections. So according to the generalized Grothendieck p-curvature conjecture they should be geometric. But here unlike for Katz, we do not know a priori the geometricity, it comes ultimately as a consequence of the more difficult finiteness result which the authors prove in *loc. cit.*

Chapter 3
Lecture 3: Malčev-Grothendieck's Theorem, Its Variants in Characteristic $p > 0$, Gieseker's Conjecture, de Jong's Conjecture, and the One to Come

Abstract We recall Malčev's theorem to the effect that the triviality of the profinite completion of a finitely generated group implies the triviality of its algebraic completion. We recall Grothendieck's version of it formulated with $\mathcal{D}$-modules using the Riemann-Hilbert correspondence, then the Gieseker conjecture, its counterpart in characteristic $p > 0$, its solution and generalizations.

3.1 Over $\mathbb{C}$

Let Γ be a finitely generated group, $\hat{\Gamma}$ be its profinite completion, Γ^{alg} be its proalgebraic completion. So $\hat{\Gamma} = \varprojlim_{\{G \to Q\}} Q$ where Q runs through all finite quotients, $\Gamma^{\text{alg}} = \varprojlim_{\{G \to Q\}} Q$ where Q runs through all algebraic subgroups of $GL_r(\mathbb{C})$ for some $r \in \mathbb{N}_{>0}$.

Theorem 3.1 (Malčev [Mal40]) $\hat{\Gamma} = \{1\} \Longrightarrow \Gamma^{\text{alg}} = \{1\}$.

Proof Indeed, if $\rho : \Gamma \to GL_r(\mathbb{C})$ is a representation, then $Q = \text{Im}(\rho)$ lies in $GL_r(A)$ where A is a ring of finite type over $\mathbb{Z}$. Let $\mathfrak{m} \subset A$ be a maximal ideal, so with finite residue field $A/\mathfrak{m}$, and define $\iota : A \hookrightarrow \hat{A} = \varprojlim_{n \in \mathbb{N}} A/\mathfrak{m}^n$ to be the completion of A at $\mathfrak{m}$. Then ρ is non-trivial if and only if $\iota \circ \rho : \Gamma \to GL_r(\hat{A})$ is non-trivial. As $\iota \circ \rho$ factors through $\hat{\rho} : \hat{\Gamma} \to GL_r(\hat{A})$, then $\hat{\Gamma} = \{1\} \Longrightarrow \rho$ is trivial, i.e. has image equal to the identity matrix $\mathbb{I}_r \in GL_r(A)$.

□

Remark 3.2 If Γ is not finitely generated, the implication of the theorem might be wrong. For example, let $\Gamma = \mathbb{Q}$, then $\hat{\Gamma} = \{1\}$, but e.g. the representations $\chi : \mathbb{Q} \to GL_1(\mathbb{C})$, $n \mapsto \exp(2\pi\sqrt{-1}an)$ for some $a \in \mathbb{C}$ or $\rho : \mathbb{Q} \to GL_2(\mathbb{C})$, $n \mapsto \begin{pmatrix} 1 & n \\ 0 & 1 \end{pmatrix}$ are non-trivial.

We now assume that X is a smooth quasi-projective variety defined over $\mathbb{C}$. Then the Riemann-Hilbert correspondence [Del70] is an equivalence between complex local systems and algebraic integrable connections on X which are regular singular at

H. Esnault, *Local Systems in Algebraic-Arithmetic Geometry*, Lecture Notes in Mathematics 2337, https://doi.org/10.1007/978-3-031-40840-3_3

infinity [Del70], or equivalently $\mathcal{O}_X$-coherent regular singular $\mathcal{D}_X$-modules. On the other hand, the topological fundamental group $\pi_1(X(\mathbb{C}), x)$ based at some complex point x is finitely generated, even finitely presented but this is not used here in the discussion, and by the Riemann Existence Theorem, its profinite completion is Grothendieck's étale fundamental group $\pi_1(X_\mathbb{C}, x_\mathbb{C})$. So Malčev's theorem can be rephrased in the context of complex algebraic geometry by saying

Theorem 3.3 (Grothendieck [Gro70]) $\pi_1(X_\mathbb{C}, x_\mathbb{C}) = \{1\} \Longrightarrow$ *there are no non-trivial $\mathcal{O}_X$-coherent regular singular $\mathcal{D}_X$-modules on X.*

3.2 Over an Algebraically Closed Field of Characteristic $p > 0$

I am not aware of a direct proof of Grothendieck's theorem without passing through the Riemann-Hilbert correspondence. Gieseker [Gie75, p.8] conjectured the "same" theorem over an algebraically closed field of characteristic $p > 0$.

Theorem 3.4 ([EM10], Theorem 1.1) *Let X be a smooth projective variety defined over an algebraically closed field k of characteristic $p > 0$. If $\pi_1(X) = \{1\}$, there are no non-trivial $\mathcal{O}_X$-coherent $\mathcal{D}_X$-modules.*

Proof Following Katz [Gie75, Theorem 1.3], we first identify $\mathcal{O}_X$-coherent $\mathcal{D}_X$-modules with Frobenius divided sheaves. Then the strategy consists, starting from a non-trivial Frobenius divided sheaf $E = E_0 = \mathrm{Frob}^* E_1, E_1 = \mathrm{Frob}^* E_2, \dots$ of rank r, to find a non-trivial vector bundle M of rank r with the property that $(\mathrm{Frob}^n)^*(M) \cong M$ for some $n \in \mathbb{N}_{>0}$. Indeed the underlying Lang torsor under $GL_r(\mathbb{F}_{p^n})$ is finite étale over X, thus has to be trivial by assumption, so M has to be trivial, a contradiction. To this aim, we observe that the Frobenius divisibility forces all kinds of Chern classes of E to be trivial and E_i itself to be stable for i large if the object $(E_0, E_1, E_2, \dots)$ itself is simple. So we consider in the moduli of vector bundles of rank r with vanishing Chern classes the Zariski closure of the locus of the E_n for n large, on which Frobenius is a rational dominant map. We can then specialize the situation to X over $\bar{\mathbb{F}}_p$ and apply Hrushovski's theorem to the effect that there are (many) preperiodic points. By the above discussion, those yield non-trivial Lang torsors of $X_{\bar{\mathbb{F}}_p}$, thus by Grothendieck's specialization homomorphism, also a non-trivial finite étale cover of X, a contradiction. □

Remarks 3.5

(1) Hrushovski's proof [Hru04], which relies on model theory, has been meanwhile proven purely in the framework of arithmetic geometry by Varshavski, see [Var18].
(2) The core of the proof is to find such an M with $(\mathrm{Frob}^n)^*(M) \cong M$ for some $n \in \mathbb{N}_{>0}$. This unfortunately does not seem to be deducible from Katz' 'Riemann-Hilbert' theorem in characteristic $p > 0$ or its (so far) ultimate generalization

[BL17]. So in the present state of understanding, we cannot draw a parallel with the complex proof.

The category of $\mathcal{O}_X$-coherent $\mathcal{D}_X$-modules is equivalent to the one of crystals in the infinitesimal site, as developed by Grothendieck. This led J. de Jong to pose the following vast strengthening of Gieseker conjecture (Theorem 3.4):

Conjecture 3.6 (de Jong 2010, see [ES18]) Let X be a smooth projective variety defined over an algebraically closed field k of characteristic $p > 0$. If $\pi_1(X) = \{1\}$, there are no non-trivial isocrystals.

There are (very) partial results on the conjecture, see notably [ES18, Theorem 1.2]. The main obstruction to our understanding lies in the prefix "iso". All approaches I know start with the mod p reduction of a crystal associated to an isocrystal and try

- to use again an argument as in the proof of Theorem 3.4 on the existence of M, with some sharpening (see [ES18, Proposition 3.2]), so ultimately it relies on Hrushovski's theorem; this also requests vanishing results for Chern classes of the mod p reduction of crystals, see [ES19], which have been generalized by Bhatt-Lurie (yet unpublished);
- so to trivialize the crystal itself (see e.g. [ES18, Corollary 3.8]), not only the isocrystal.

In rank 1 it is irrelevant (see [ES18, Lemma 2.12]), not however in higher rank. One would wish to find a Frobenius invariant isocrystal on the scratch, without passing through randomly chosen underlying crystals. However Hrushovski's theorem is no longer of help.

3.3 Over a p-Adic Field

The wish would be to have a version of de Jong's conjecture for prismatic "iso"-crystals. Here one problem is that the notion of prismatic crystals is documented [BS21], not however the one of prismatic isocrystals. One question is what one has to invert to make the problem meaningful. The formulation should also be compatible with a complex embedding of the p-adic field and the initial Grothendieck formulation over $\mathbb{C}$.

Chapter 4
Lecture 4: Interlude on Some Similarity Between the Fundamental Groups in Characteristic 0 and $p > 0$

Abstract We describe two theorems concerning the structure of the (tame) fundamental group of a smooth (quasi)-projective variety in characteristic $p > 0$. The first one (Esnault et al., Sel Math 28(2):19 pp., 2022) shows a similarity with the complex situation concerning the finite presentation and ultimately relies on a theorem of Lubotzky (J. Algebra 242(2):672–690, 2001), the second one (Esnault et al., Algebr. Geom. 10(3):327–347, 2023) yields a new obstruction for a smooth projective variety in characteristic $p > 0$ to lift to characteristic 0 and ultimately relies on Grothendieck's specialization homomorphism. In particular, and interestingly, the second one shows that we cannot explain the finite presentation of the first one by a lifting argument. In Chap. 4 we treat the finite presentation, in Chap. 5 the obstruction.

4.1 Lubotzky's Theorem

Definition 4.1 ([Lub01], p. 1 of the Introduction) A profinite group π is said to be

(1) finitely generated if there is a free profinite group F on finitely many generators and a continuous surjection $F \xrightarrow{\epsilon} \pi$;
(2) finitely presented if it is finitely generated and $\mathrm{Ker}(\epsilon)$ is finitely generated as a closed normal subgroup of F.

Concretely, in (1) there are finitely many elements $f_i, i = 1, \ldots, N$ in F such that the abstract subgroup $\langle f_1, \ldots, f_N \rangle \subset F$ surjects to any finite quotient $\pi \to G$ in the profinite structure of π. For 2) there are finitely many elements $r_j, j = 1, \ldots, M$ in F such that $\mathrm{Ker}(F \to \pi)$ is spanned by $\langle f r_j f^{-1},\ j = 1, \ldots, M, f \in F \rangle$. Finite generation and presentation are well defined, they do not depend on the chosen presentation. For the finite generation, it is very classical, for the finite presentation, a slick way is simply to apply the following theorem.

Theorem 4.2 (Lubotzky [Lub01], Theorem 0.3) *π is finitely presented if and only if it is finitely generated and in addition there is a constant $C \in \mathbb{R}_{>0}$, such that for*

H. Esnault, *Local Systems in Algebraic-Arithmetic Geometry*, Lecture Notes in Mathematics 2337, https://doi.org/10.1007/978-3-031-40840-3_4

any $r \in \mathbb{N}_{>0}$, *for any prime number* ℓ, *for any continuous linear representation* $\rho : \pi \to GL_r(\mathbb{F}_\ell)$, *it holds*

$$\dim_{\mathbb{F}_\ell} H^2(\pi, \rho) \leq C \cdot r.$$

Example 4.3 The standard example is given by π being the profinite completion of an abstract finitely presented group, in the classical sense, so with the same definition without topology.

4.2 Tame Fundamental Group

Let X be a regular connected scheme. For x a closed point of the normal compactification $\bar{X}$ of X, we denote by $k(X)_x$ the field of fractions of the completed local ring of $\bar{X}$ at x. For a connected finite étale cover $\pi : Y \to X$, we denote by $\bar{Y}$ the normalization of $\bar{X}$ in the field of functions on Y. Any closed point $x \in \bar{X}$ defines for any closed point $y \in \bar{Y}$ above x a finite field extension $k(X)_x \hookrightarrow k(Y)_y$.

Definition 4.4 ([KS10], Introduction and Theorem 1.1)

(1) If X has dimension 1, a finite étale cover $\pi : Y \to X$ is tame if and only if all the field extensions $k(X)_x \hookrightarrow k(Y)_y$ are tame, that is the indexes of ramification are prime to p and the residual extension $k(x) \hookrightarrow k(y)$ are separable.
(2) In higher dimension, π is tame if and only if for any morphism of a normal dimension 1 scheme $C \to X$, the induced morphism $C \times_X Y \to C$ is tame on all the irreducible components.

4.3 Grothendieck's Specialization Homomorphism, [SGA1], Exposé X, Exposé XIII

Let $X_S \to S$ be a smooth morphism, where S is any scheme. We consider two field value points $\mathrm{Spec}(F) \to S$ and $\mathrm{Spec}(k) \xrightarrow{s} S$ with the property that $\mathrm{Spec}(k)$ lies in the Zariski closure of $\mathrm{Spec}(F)$. So there is an irreducible subscheme $Z \subset S$ such that $s \in Z$ is a point and $\mathcal{O}(Z) \to F$ is injective. Let $\widehat{Z_s}$ is the completion of Z at s, and $F \hookrightarrow \hat{F}_s$ a field extension such that $\hat{F}_s$ contains $\mathcal{O}(\widehat{Z_s})$.

If X is not proper, we assume in addition that it admits a compactification $X_S \hookrightarrow \bar{X}_S$ so $\bar{X}_S \to S$ is smooth proper, and $\bar{X}_S \setminus X_S \to S$ is a relative normal crossings divisor with smooth components. We call it a *good* compactification (over S). So we have a diagram

$$\mathrm{Spec}(\hat{F}_s) \longrightarrow \widehat{Z_s} \longleftarrow s$$

together with the scheme over it

$$\begin{array}{ccccc} X_{\hat{F}_s} & \longrightarrow & X_{\widehat{Z_s}} & \longleftarrow & X_s \\ \downarrow & & \downarrow & & \downarrow \\ \mathrm{Spec}(\hat{F}_s) & \longrightarrow & \widehat{Z_s} & \longleftarrow & s \end{array}$$

We denote by $\overline{\hat{F}}_s \supset \hat{F}_s$ and $\bar{k} \supset k$ algebraic closures, the latter defining $\bar{s} \to s$. Then, upon choosing an S-point $x_S : S \to X_S$, one defines [SGA1, Exposé XIII 2.10] a specialization homomorphism

$$sp_{\hat{F}_s,s} : \pi_1^t(X_{\hat{F}_s}, x_{\hat{F}_s}) \to \pi_1^t(X_s, x_s)$$

which is the composite of the functoriality homomorphism

$$\pi_1^t(X_{\hat{F}_s}, x_{\hat{F}_s}) \to \pi_1^t(X_{\widehat{Z_s}}, x_s)$$

with the inverse of the base change isomorphism

$$\pi_1^t(X_s) \xrightarrow{\cong} \pi_1^t(X_{\widehat{Z_s}}).$$

Finally one has the functoriality homomorphism

$$\pi_1^t(X_{\hat{F}_s}, x_{\hat{F}_s}) \to \pi_1^t(X_F, x_F)$$

which is an isomorphism in restriction to the geometric fundamental groups

$$\pi_1^t(X_{\overline{\hat{F}}_s}, x_{\overline{\hat{F}}_s}) \xrightarrow{\cong} \pi_1^t(X_{\bar{F}}, x_{\bar{F}}).$$

Taken together this defines the specialization homomorphism

$$sp_{F,s} : \pi_1^t(X_F, x_F) \to \pi_1^t(X_s, x_s)$$

which, while restricted to the geometric fundamental groups, defines the specialization homomorphism

$$sp_{\bar{F},\bar{s}} : \pi_1^t(X_{\bar{F}}, x_{\bar{F}}) \to \pi_1^t(X_{\bar{s}}, x_{\bar{s}}).$$

The specialization homomorphisms $sp_{F,s}$ and $sp_{\bar{F},\bar{s}}$ are surjective, and $sp_{\bar{F},\bar{s}}$ induces an isomorphism on the pro-p'-completion [SGA1, Exposé XIII 2.10, Corollaire 2.12].

4.4 Finite Generation

We apply Grothendieck's specialization homomorphism in the following situation. Let X be smooth projective defined over a characteristic 0 field together with a good compactification $X \hookrightarrow \bar{X}$, let $X_S \hookrightarrow \bar{X}_S$ be a good compactification over S as in Sect. 4.3. Upon choosing a S-point $x_S : S \to X_S$, we have the specialization homomorphism

$$sp_{F,s} : \pi_1(X_F, x_F) \to \pi_1^t(X_s, x_s)$$

where F is a geometric point of $X_{k(S)}$ which specializes to a $\bar{\mathbb{F}}_p$-point s of S. (Beware we slightly change the notation as compared to Sect. 4.3 as we consider only the geometric fundamental groups). Taking $F = \mathbb{C}$ then $\pi_1(X_F, x_F)$ is by Riemann existence theorem [SGA1, Exposé XII, Théorème 5.1] the profinite completion of $\pi_1(X(\mathbb{C}), x(\mathbb{C}))$, the latter being itself an *abstract finitely presented group*, see [Esn17, Theorem 1.1]. Thus $\pi_1^t(X_s, x_s)$ is *finitely generated.*

Theorem 4.5 ([ESS22], Theorem 1.1) *Let X be a smooth quasi-projective variety defined over an algebraically closed field k of characteristic $p > 0$, with a good compactification $X \hookrightarrow \bar{X}$. Then $\pi_1^t(X)$ (based at any geometric point x) is finitely presented.*

Remark 4.6

(1) The tameness assumption is necessary. If X is not proper, "in the rule" $\pi_1(X)$ is not finitely generated. For example, for each natural number s, and a choice of s natural numbers m_i prime to p and pairwise different, the Artin-Schreier covers $y^p - y = z^{m_i}$ of the affine line $\mathbb{A}_k^1 = \mathrm{Spec}(k[z])$ yield a surjection $\pi_1(\mathbb{A}_k^1) \twoheadrightarrow \oplus_{i=1}^s \mathbb{Z}/p\mathbb{Z}$.
(2) We need the good compactification assumption in the argument but it should not be necessary as the result does not see the good compactification. In addition we (perhaps?) believe in resolution of singularities. We'll mention in the proof which replacement for resolution of singularities we need, which might be called a "numerical resolution". Hübner and Temkin announced a solution to the problem.

4.5 Proof of Theorem 4.5, the $\ell \neq p$ Part

We could call this part of the proof a "SGA type proof". All the ingredients are contained there. We have to prove the linear growth in r of $H^2(\pi_1^t, \rho)$ for a representation

$$\rho : \pi_1^t \to GL_r(\mathbb{F}_\ell)$$

(which is necessarily continuous). Here we simplify the notation $\pi_1^t(X, x)$ to π_1^t and $\pi_1(X, x)$ to π_1. We denote by $X^\rho \to X$ the cover defined by Ker(ρ) which by assumption factors the universal tame cover $X^t \to X$. We split the universal cover based at x

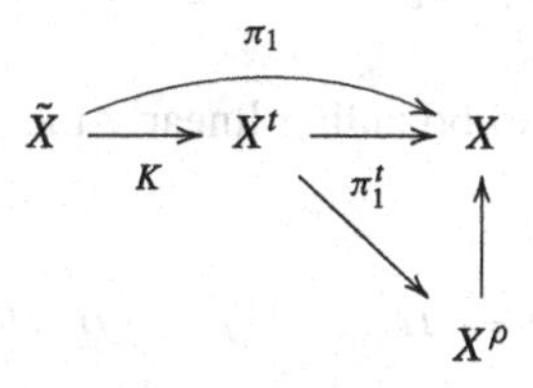

We denote by M the underlying vector space $\oplus_1^r \mathbb{F}_\ell$ of the ρ-representation. The Hochshild-Serre spectral sequence yields the exact sequence

$$(H^1(K, \mathbb{F}_\ell) \otimes M)^{\pi_1^t} \to H^2(\pi_1^t, M) \to H^2(\pi_1, M).$$

By definition there is no $\mathbb{F}_\ell$-abelian quotient of K, thus

$$(\star) \quad 0 = H^1(K, \mathbb{F}_\ell) = H^1(K^{ab}, \mathbb{F}_\ell),$$

from which we derive $H^2(\pi_1^t, M) \hookrightarrow H^1(\pi_1, M)$. On the other hand, the Hochshild-Serre spectral sequence induces the exact sequence

$$(H^1(\tilde{X}, \mathbb{F}_\ell) \otimes M)^{\pi_1} \to H^2(\pi_1, M) \to H^2(X, M)$$

and $H^1(\tilde{X}, \mathbb{F}_\ell) = 0$. So $H^2(\pi_1, M) \hookrightarrow H^1(X, M)$ is injective as well. Taken together this yields that the composed linear map

$$H^2(\pi_1^t, M) \hookrightarrow H^2(\pi_1, M) \hookrightarrow H^2(X, M)$$

is injective. We show the existence of the linear bound on the a priori larger group $H^2(X, M)$.

By the Lefschetz theorem [EK16, Theorem 1.1 b)] we may assume that X is a surface. Using an "ample" curve $C \to X$ in good position so the compactification of C and the boundary of X form a strict normal crossings divisor and $X \setminus C$ is affine, purity yields the exact sequence

$$H^0(C, M) \to H^2(X, M) \to H^2(X \setminus C, M).$$

As $\dim_{\mathbb{F}_\ell} H^0(C, M) \leq r$, we thus may assume that X is affine. On the other hand, by Deligne's theorem [Ill81, Corollaire 2.7], tameness implies

$$\chi(X, \mathbb{M}) = r\chi(X, \mathbb{F}_\ell)$$

and we always have

$$\chi(X, \mathbb{F}_\ell) = \chi(X, \mathbb{Q}_\ell).$$

The latter group is independent of $\ell \neq p$ by Deligne's purity [Del80, Théorème 3.3.1].

This reduces the problem to bounding linearly the growth of $H^1(X, M)$. From the exact sequence

$$0 \to H^1(\pi_1^t, M) \to H^1(\pi, M) \to (H^1(K, \mathbb{F}_\ell) \otimes M)^{\pi_1^t},$$

again using ($\star$), we see that $H^1(X, M)$, which is equal to $H^1(\pi_1, M)$, is equal to $H^1(\pi_1^t, M)$. As a 1-cocycle $\varphi : \pi_1^t \to M$ fulfils $\varphi(ab) = a\varphi(b) + \varphi(a)$, it holds

$$\dim_{\mathbb{F}_\ell} H^1(\pi_1^t, M) \leq \delta r$$

where δ is the number of topological generators of π_1^t. This finishes this computation, which in fact can be performed even if we do not have a good model, using alterations à la de Jong-Gabber-Temkin.

4.6 Proof of Theorem 4.5, the p Part

We denote by $j : X \hookrightarrow \bar{X}$ the good compactification. We denote by $\underline{M}$ the local system associated to M. The key point is to relate $H^2(\pi_1^t, M)$ to a cohomology group which comes from cohomology of some extension on $\bar{X}$ of the local system $\underline{M}$ on X. Being able to use local systems rather than abstract representations yields more flexibility as we have at disposal the cohomology of all kinds of constructible extensions of the local systems as opposed to just group cohomology.

Theorem 4.7 *There is an injective $\mathbb{F}_p$-linear map*

$$H^2(\pi_1^t(X), M) \to H^2(\bar{X}, j_* \underline{M}).$$

Proof First let us fix what we have to prove: it holds

$$H^2(\pi_1^t, M) = \varinjlim_{U \subset \pi^t \text{open normal}} H^2(G_U, M^U), \; G_U = \pi_1^t/U.$$

For each such U we define the fibre square

$$\begin{array}{ccc} X_U & \xrightarrow{j_U} & \bar{X}_U \\ h_U \downarrow & & \downarrow \bar{h}_U \\ X & \xrightarrow{j} & \bar{X} \end{array}$$

where h_U is the tame Galois cover defined by G_U and $\bar{X}_U$ is the normalization of $\bar{X}$ in the field of functions $k(X_U)$ of X_U. As j is a normal crossings compactification, $\bar{h}_U$ is *numerically tame* in the sense of [KS10, Section 5], see [KS10, Theorem 5.4 (a)]. Thus by [Gro57, Corollaire p.204] one has a simple way to compute the equivariant cohomology groups $H^n(\bar{X}, j_*\underline{M})$:

$$R^{>0}\bar{h}_{U*}^{G_U} j_{U*}\underline{M}^U = 0,$$

$$H^n(\bar{X}_U, G_U, j_{U*}\underline{M}^U) = H^n(\bar{X}, j_*\underline{M}).$$

We conclude that the spectral sequence converging to equivariant cohomology

$$E_2^{ab} = H^a(G_U, H^b(\bar{X}_U, j_{U*}\underline{M}^U)) \Rightarrow H^{a+b}(\bar{X}_U, G_U, j_{U*}\underline{M}^U),$$

for each $U \subset \mathrm{Ker}(\rho)$ yields a short exact sequence

$$H^0(G_U, H^1(\bar{X}_U, j_{U*}\underline{M}^U)) \to H^2(G_U, H^0(\bar{X}_U, j_{U*}\underline{M}^U)) \to H^2(\bar{X}, j_*\underline{M})$$

where

$$H^1(\bar{X}_U, j_{U*}\underline{M}^U)^{G_U} = (H^1(\bar{X}_U, \mathbb{F}_p) \otimes_{\mathbb{F}_p} M)^{G_U}.$$

On the other hand, $H^1(\bar{X}_U, \mathbb{F}_p) \to H^1(X_U, \mathbb{F}_p)$ is injective and again by $(\star)$ it holds $\lim_U H^1(\bar{X}_U, \mathbb{F}_p) = 0$.

□

Remark 4.8 We see that the main point is that a good normal crossings compactification implies that the tower h_U is numerically tame. The problem implicitly posed is the following: how, out of a random normal compactification, can we construct one such that for $U \subset \pi^t$ open normal, h_U is numerically tame.

Once there, we can cut down again by a Lefschetz type argument [EK16, Theorem 1.1 b)] to surfaces and on them, using the Artin-Schreier exact sequence

$$0 \to j_*\underline{M} \to \mathcal{M} \xrightarrow{1-F} \mathcal{M} \to 0$$

where $\mathcal{M}$ is a locally free sheaf on $\bar{X}$ with restriction to X equal to $\underline{M} \otimes_{\mathbb{F}_p} \mathcal{O}_X$, and the classical fact that this sequence remains exact on cohomology, we are reduced to bounding above $\dim_k H^2(\bar{X}, \mathcal{M})$ linearly in r, which is the same as $\dim_k H^0(\bar{X}, \mathcal{M}^\vee \otimes \omega_{\bar{X}})$. We are back to cohomology of coherent sheaves.

Now the main point is the following. If $h : Y \to X$ is the Galois cover defined by the monodromy representation of $\underline{M}$, with Galois group G, thus $h^*\underline{M}$ is trivial. Let $\bar{h} : \bar{Y} \to \bar{X}$ be its compactification as above, then $\mathcal{M} = (\bar{h}_*\mathcal{L})^G$, so $\bar{h}^*\mathcal{M} \subset \mathcal{L} := H^0(Y, h^*\underline{M}) \otimes_{\mathbb{F}_p} \mathcal{O}_{\bar{Y}}$ and as a consequence of Abhyankar's lemma, tameness implies $\mathcal{L} \otimes \bar{h}^*\mathcal{O}_{\bar{X}}(-D) \subset \bar{h}^*\mathcal{M}$ for $D = (\bar{X} \setminus X)_{\mathrm{red}}$. Those two informations together yield

$$\mathcal{L} \otimes \bar{h}^*\mathcal{O}_{\bar{X}}(-D) \subset \bar{h}^*\mathcal{M} \subset \mathcal{L}$$

which is a boundedness statement. This enables us to conclude that the restriction map

$$H^0(\bar{X}, \mathcal{M}^\vee \otimes \omega_{\bar{X}}) \to H^0(C \cap C', \mathcal{M}^\vee \otimes \omega_{\bar{X}})$$

is injective for C, C' generic curves in the linear system of $\mathcal{H}' = \mathcal{H} \otimes \omega_{\bar{X}}(D)$, where $\mathcal{A}$ is chosen to be very ample and we request $\mathcal{H}'$ to be very ample as well. Note $\mathcal{H}$ and $\mathcal{H}'$ depend only on X and r. On the other hand, $C \cap C'$ is the union of $c_2(\mathcal{H}')$-points so the right hand side of the inequality is equal to $c_1(\mathcal{H}')^2 \cdot r$. This finishes the proof.

Chapter 5
Lecture 5: Interlude on Some Difference Between the Fundamental Groups in Characteristic 0 and $p > 0$

Abstract See the Abstract of Chap. 4: we show here the existence of an obstruction to lift a smooth (quasi-)projective variety defined over an algebraically closed field k of characteristic $p > 0$ to characteristic 0 which relies purely on the shape of its (tame) fundamental group.

5.1 Serre's Construction

This problem has been addressed for the first time by Serre [Ser61]. We use the notation of Sect. 4.2 but assume more concretely that k is an algebraically closed field of characteristic $p > 0$, $X_k =: X$ is smooth projective and obtained as follows. There is a finite Galois étale cover $Y \to X$ such that $Y \hookrightarrow \mathbb{P}^n$ is a smooth complete intersection of dimension ≥ 3. So in particular $\pi_1(Y) = \{1\}$. In addition, the Galois group G is the restriction of a linear action $\rho : G \to GL_{n+1}(k)$. Then [Ser61, Lemma] shows that if X lifts to X_R with $S = \mathrm{Spec}(R)$, R a noetherian local complete ring, then ρ lifts to $\rho_R : G \to PGL_{n+1}(R)$, which is not possible if the cardinality of the p-Sylow subgroups of G is large.

5.2 Various Obstructions

We mention three major directions of obstructions which have been settled since Serre's work. Clearly this list is not exhaustive.

Deligne-Illusie in [DI87] proved that X smooth proper, lifting to $W_2(k)$, k perfect of characteristic $p > 0$ with $p > \dim(X)$, has the property that its Hodge to de Rham spectral sequence degenerates in E_1. This yields an obstruction to lift to $W_2(k)$ as examples for which the spectral sequence does not degenerate were previously known [Ray78]. This has been the basis of vast further developments.

H. Esnault, *Local Systems in Algebraic-Arithmetic Geometry*, Lecture Notes in Mathematics 2337, https://doi.org/10.1007/978-3-031-40840-3_5

Achinger-Zdanowicz construct in [AZ17] specific varieties which are non-liftable to characteristic 0 by blowing up the graph of Frobenius which is assumed to be non-liftable in a rigid variety with no corner piece of the F-filtration. It is remarkable that their example has cohomology of Tate type.

van Dobben de Bruyn proved that if $X \subset C^3$ is a smooth ample divisor where C is a supersingular genus ≥ 2 curve over $\bar{\mathbb{F}}_p$, then X does not lift to characteristic 0, nor does any smooth proper variety which is rationally dominant over X, [vDdB21, Theorem. 1]. The main property [vDdB21, Theorem. 2] is that if $X_S \to S$ lifts X, and X admits a morphism to a smooth projective genus ≥ 2 curve C, then after possibly base changing with an inseparable cover of C, the morphism lifts to characteristic 0.

5.3 An Abstract Obstruction to Lift to Characteristic 0, Based on the Structure of the Fundamental Group

The Hodge-de Rham obstruction singled out by Deligne-Illusie is of theoretical nature, that is it does not depend on a concrete way to construct the variety. However it is *not* an obstruction to lift to characteristic 0; there are schemes which lift to characteristic 0, yet in a ramified way, not over W_2. A classical example is a supersingular Enriques surface over k algebraically closed of characteristic 2, see [Ill79, Proposition II.7.3.8]. The other obstructions to lift to characteristic 0 rely on the construction of the variety.

The aim of the remaining part of Chap. 5 is to show that there is an essential difference between the prime to p quotient of the fundamental group of varieties in characteristic $p > 0$ and the one in characteristic 0. It provides a *conceptual* obstruction. It is in *contrast* with the similarity we explained in Chap. 4, where we showed that the (tame) fundamental group of a smooth (quasi-)projective variety defined over an algebraically closed field of characteristic $p > 0$ (admitting a good compactification) is finitely presented, as it is in characteristic 0. It is also in contrast with the foundational theorem by Achinger [Ach17, Theorem 1.1.1] after which every connected affine scheme of positive characteristic is a $K(\pi, 1)$ space for the étale topology. His theorem notably says that affine varieties over an algebraically closed field in characteristic $p > 0$ are analog to Artin neighborhoods in characteristic 0, see [SGA4, Exposé XI]. The theorem means precisely that for any locally constant étale sheaf of finite abelian groups $\mathcal{F}$ on X, the homomorphisms

$$H^i(\pi_1(X, x), \mathcal{F}_x) \to H^i(X, \mathcal{F})$$

coming from the Hochshild-serre spectral sequence are isomorphisms for all i. Here $x \to X$ is a geometric point. It has not really been documented in the literature, but we could think of Achinger's theorem as the building block of the theory of étale cohomology, reducing it to continuous group cohomology. In this direction, we mention [BGH18, Theorem 13.0.1], a reference kindly provided by one of the referees.

5.4 Main Definition

Definition 5.1 (See [ESS22b], Definition A) A profinite group π is said to be p'-discretely finitely generated (resp. p'-discretely finitely presented) if there is a finitely generated (resp. presented) discrete group Γ together with a group homomorphism $\gamma : \Gamma \to \pi$ such that

(1) the profinite completion $\hat{\gamma} : \hat{\Gamma} \to \pi$ is surjective;
(2) for any open subgroup $U \subset \pi$ with $\Gamma_U := \gamma^{-1}(U)$ the restriction $\gamma_U : \Gamma_U \to U$ induces a continuous group isomorphism on pro-p'-completions

$$\gamma_U^{(p')} : \Gamma_U^{(p')} \to U^{(p')}.$$

Grothendieck's specialization homomorphism 4.3 together with the lifting property [EGAIV$_4$, Théorème 18.1.2] imply that if $\pi = \pi_1(X, x)$ is the fundamental group of a smooth proper variety X, then it is p'-discretely finitely presented (in particular it is p'-discretely finitely generated), see [ESS22b, Proposition 2.7]. Property (1) is also true for $\pi_1^t(X, x)$ when there is a good compactification which comes from characteristic 0. But to check it in the tower as requested in (2) is more subtle. We thank the referee of [ESS22b] for kindly noticing this. Nonetheless the property is true, see [ESS22b, Example 2.8].

Example 5.2 Of course if as in Example 4.3, π is the profinite completion of a finitely presented (resp. generated) group, then π is p'-discretely finitely presented (resp. generated).

Theorem 5.3 ([ESS22b], Theorem C) *There are smooth projective varieties X defined over an algebraically closed field k of characteristic $p > 0$ such that $\pi_1(X, x)$ is not p'-discretely finitely generated. In particular this notion is an obstruction to liftability to characteristic* 0.

Remark 5.4 We remark that in view of Example 5.2 and of the finite presentation of Theorem 4.5, Theorem 5.3 implies in particular that there are smooth projective varieties X defined over an algebraically closed field k of characteristic $p > 0$ such that $\pi_1(X, x)$ is not the profinite completion of a finitely presented group. In fact, this property is easier to see than Theorem 5.3 itself.

5.5 Independence of ℓ and Schur Rationality

Let π be a profinite group, and let $\varphi : \pi \twoheadrightarrow G$ be a continuous finite quotient with kernel $U_\varphi = \mathrm{Ker}(\varphi)$. We denote by U_φ^{ab} its abelianization. Then conjugation induces a commutative diagram

$$\begin{array}{ccccc} \pi & \longrightarrow & \mathrm{Aut}(U_\varphi) & \longrightarrow & \mathrm{Aut}(U_\varphi^{ab}) \\ \downarrow \varphi & & \downarrow & \nearrow & \\ G & \longrightarrow & \mathrm{Out}(U_\varphi) & & \end{array}$$

If π is finitely generated, then U_φ is finitely generated so U_φ^{ab} is a finitely generated $\hat{\mathbb{Z}}$-module. We set

$$\rho_{\varphi,\ell} : G \to \mathrm{GL}(U_\varphi^{ab} \otimes \mathbb{Q}_\ell),$$

for the induced representation, with character

$$\chi_{\varphi,\ell} = \mathrm{Tr}(\rho_{\varphi,\ell}) : G \to \mathbb{Q}_\ell.$$

The first main point is the following proposition.

Proposition 5.5 (See [ESS22b], Propositions 3.4, 3.5)

(1) If π is p'-discretely finitely generated, then for all $\ell \neq p$, $\chi_{\varphi,\ell}$ has values in $\mathbb{Z}$ and is independent of ℓ.

(2) If X is a smooth projective variety defined over an algebraically closed characteristic $p > 0$ field, and $\varphi : \pi \twoheadrightarrow G$ is a finite quotient (thus in particular φ is continuous), then for all $\ell \neq p$, $\chi_{\varphi,\ell}$ has values in $\mathbb{Z}$ and is independent of ℓ.

Proof The property (1) comes essentially from the definition: setting

$$\Gamma_\varphi = \mathrm{Ker}(\Gamma(\to \pi) \to G)$$

we have for $\ell \neq p$ the relation

$$\Gamma_\varphi^{ab} \otimes_{\mathbb{Z}} \mathbb{Q}_\ell = U_\varphi^{ab} \otimes_{\mathbb{Z}} \mathbb{Q}_\ell.$$

The property (2) is more interesting in view of the final result, see Theorem 5.10. Let $Y \to X$ be the Galois cover with group G. Then

$$U_\varphi^{ab} \otimes_{\mathbb{Z}} \mathbb{Q}_\ell = H^1(Y, \mathbb{Q}_\ell)^\vee.$$

As a consequence of the Weil conjectures, see [KM74], the characteristic polynomial of $g \in G$ acting on $H^1(Y, \mathbb{Q}_\ell)$ lies in $\mathbb{Z}[T]$ and does not depend on ℓ.

□

Remark 5.6 Part (1) of the proposition makes clear that the independence of $\ell \neq p$ property is shared by all smooth projective varieties defined over an algebraically closed field of characteristic $p > 0$, which in addition lift to characteristic 0. Indeed then their fundamental group is p'-discretely finitely presented. Part (2) shows that in fact the independence of $\ell \neq p$ property is shared by all smooth projective varieties defined over an algebraically closed field of characteristic $p > 0$. So the property is not an obstruction to liftability. On the contrary, we shall use it now in order to define our sought obstruction, which shall be a rationality obstruction.

The second main point is the following Proposition.

Proposition 5.7 (See [ESS22b], Proposition 3.7) *If π is p'-discretely finitely generated, then for any continuous finite quotient $\varphi : \pi \twoheadrightarrow G$, there is a $\mathbb{Q}$-vector space V_φ and a representation $\rho_\varphi : G \to GL(V_\varphi)$ such that for every $\ell \neq p$, the relation*

$$\rho_{\varphi,\ell} = \rho_\varphi \otimes_{\mathbb{Q}} \mathbb{Q}_\ell$$

holds.

Proof Indeed, $V_\varphi = \Gamma_\varphi^{ab}$ and $\rho_\varphi : G \to GL(\Gamma_\varphi^{ab} \otimes \mathbb{Q})$.

□

We turn this into a definition.

Definition 5.8 (See [ESS22], Definition 3.6) A profinite group π satisfying independence of ℓ with the exception of p is Schur rational if for all finite continuous quotients $\varphi : \pi \to G$, there is a G-representation in a $\mathbb{Q}$ -vector space V_φ such that for all $\ell \neq p$, the ℓ-adic representations $U_\varphi^{\mathrm{ab}} \otimes_{\mathbb{Z}} \mathbb{Q}_\ell$ and $V_\varphi \otimes_Q \mathbb{Q}_\ell$ are isomorphic.

Remark 5.9 In fact, as we see in the proof, $(V_\varphi, \rho_\varphi)$ has even an integral structure, so is *Schur integral*, but our obstruction shall disregard this integrality property.

We now show the following theorem.

Theorem 5.10 *The Schur rationality is an obstruction for a smooth projective variety defined over an algebraically closed field of characteristic $p > 0$ to lift to characteristic* 0.

So we have to exhibit an example of a smooth projective variety X defined over an algebraically closed field k of characteristic $p > 0$ such that π is not Schur rational: we want a quotient $\varphi : \pi \twoheadrightarrow G$ such that there is no rational V_φ as in Definition 5.8.

5.6 The Roquette Curve, Combined with Serre's Construction

The Roquette curve is the smooth projective curve C over $\mathbb{F}_p$ for $p \geq 3$ which is the normal compactification of the affine curve with equation

$$y^2 = x^p - x.$$

It is defined in [Roq70], has genus $g = (p-1)/2$, is supersingular, and is the only curve for $p \geq 5$, (so $g \geq 2$ and ρ_ℓ is faithful), and $p > g+1$ with the property that the cardinality of its group of automorphisms G is larger than the Hurwitz bound $84(g-1)$. Precisely it is $2p(p^2-1)$ and all automorphisms are defined over $\mathbb{F}_{p^2}$. The equation of C presents it as an Artin-Schreier cover of $\mathbb{P}^1 \setminus \mathbb{P}^1(\mathbb{F}_p)$. This realizes $\mathbb{Z}/p =: N$ as a subgroup of G, which thus is a p-Sylow which is in fact normal. It is not difficult to compute (see [ESS22b, Appendix A])) that $\rho_\ell|_N$ is non-trivial, thus ρ_ℓ is absolutely irreducible. The action of $\mathbb{Q}_\ell[G] \to \mathrm{End}(H^1(C, \mathbb{Q}_\ell))$ has values in $\mathbb{Q}_\ell[G] \to \mathrm{End}_{\mathrm{Frob}}(H^1(C, \mathbb{Q}_\ell))$, and is surjective by the absolute irreducibility. The Tate conjecture identifies it with the action $\mathbb{Q}_\ell[G] \to \mathrm{End}^0(C) \otimes \mathbb{Q}_\ell$. The quaternion algebra $\mathrm{End}^0(C)$ is ramified at p and ∞, which prevents $\rho_{\varphi,\ell}$ to be rational, see [ESS22b, Proposition 4.6].

Proof of Theorem 5.10 We now take a smooth connected projective variety P defined over $\mathbb{F}_p$ of dimension at least 3, so P is simply connected over $\bar{\mathbb{F}}_p$, and on which G acts without fixpoints. We do Serre's construction setting $X = (C \times_{\mathbb{F}_p} P)/G$ where G acts diagonally. It yields an exact sequence

$$1 \to \pi_1(C) \to \pi_1(X) \xrightarrow{\varphi} G \to 1$$

which can be understood as the Galois sequence for the Galois étale cover $C \times_{\mathbb{F}_p} P \to X$ or equivalently as the homotopy exact sequence for $X \to P/G$. The action of G on $H^1(C \times_{\bar{\mathbb{F}}_p} P, \mathbb{Q}_\ell) = H^1(C_{\bar{\mathbb{F}}_q}, \mathbb{Q}_\ell)$ is the same as the outer action studied above. So it is not Schur rational. Consequently X does not lift to characteristic 0. This finishes the proof.

□

Chapter 6
Lecture 6: Density of Special Loci

Abstract By a theorem of Clemens and Landman, see Griffiths (Bull Am Math Soc 76:228–296, 1970, Theorem. 3.1) in complex geometry and Grothendieck (Séminaire de Géométrie Algébrique: Groupes de monodromie en géométrie algébrique, XIV 1.1.10, 1973) in arithmetic geometry, geometric (complex or ℓ-adic) local systems have quasi-unipotent monodromies at infinity. We explain in this section why this property is good for going from complex models to models over finite fields, and why in the Betti moduli space the complex local systems with quasi-unipotent monodromy at infinity are Zariski dense. Further, we report on Biswas et al. (Geom Topol 26(2):679–719, 2022), Landesman and Litt (Geometric local systems on very general curves and isomonodromy), Landesman and Litt (Canonical representations of surface groups) showing that the arithmetic local systems on geometric generic curves in low rank cannot be Zariski dense in their Betti moduli, contrary to what was expected in Esnault and Kerz (Camb J Math 8(3):453–478, 2020, Question 9.1 (1)) and Esnault and Kerz (Israel J Math, 9 pp., Conjecture 1.1, to appear). Finally we report on the concept of *weakly arithmetic local systems* defined in de Jong and Esnault (Trans AMS, 18 pp., Section 3, to appear), which in particular have quasi-unipotent monodromies at infinity, and show that they are Zariski dense in their Betti moduli. So for certain problems (to be defined) one would then wish to follow Drinfeld's method (Drinfeld, Math Res Lett, 8(5–6):713–728, 2001) to conclude that it is enough to check them on weakly arithmetic local systems.

6.1 Definitions

For the definition we only need X to be a normal variety and $X \hookrightarrow \bar{X}$ to be a normal compactification. This defines the codimension 1 components D_i in $\bar{X} \setminus X$ and small loops T_i around there. A representation

$$\rho : \pi_1(X, x) \to GL_r(\mathbb{C})$$

H. Esnault, *Local Systems in Algebraic-Arithmetic Geometry*, Lecture Notes in Mathematics 2337, https://doi.org/10.1007/978-3-031-40840-3_6

has *quasi-unipotent monodromies at infinity* if $\rho(T_i)$ is quasi-unipotent. This property does not depend on the choice of the compactification [Kas81, Thm.3.1]. See [EG21, Section 3] where the concept is used also for GL_r replaced by any linear algebraic group as well.

Set $\pi = \pi_1(X(\mathbb{C}), x(\mathbb{C}))$ for the topological fundamental group based at some complex point. The *framed character variety* $Ch(\pi, r)^{\square}$ is defined to be the affine variety defined over $\mathbb{Z}$ by the moduli functor which takes affine rings R to the set $\mathrm{Hom}(\pi, GL_r(R))$. It is a fine moduli functor and the resulting scheme is also called the *framed Betti moduli* $M_B(X, r)^{\square}$, also defined over $\mathbb{Z}$. The group GL_r acts by conjugation (gauge transformations) on $Ch(\pi, r)^{\square}$. Its categorial quotient

$$Ch(\pi, r) = Ch(\pi, r) /\!\!/ GL_{r,\mathbb{C}}$$

defined by

$$\mathcal{O}(Ch(\pi, r)) = \mathcal{O}(Ch(\pi, r)^{\square})^{GL_r}$$

is the *character variety*, also called the *Betti moduli space* $M_B(X, r)$. Its complex points are isomorphism classes of semi-simple local systems of rank r. The fibres of $M_B(X, r)^{\square} \to M_B(X, r)$ are the closures of GL_r-orbits. Such an orbit is closed over an irreducible complex local system.

6.2 Why Quasi-unipotent Monodromies at Infinity

The first reason is that *geometric* local systems have quasi-unipotent monodromies at infinity, see [Gri70, Theorem 3.1]. We indicate Brieskorn's complex proof [Del70, p. 125]: by base change we may assume that $Y \xrightarrow{g} U \hookrightarrow X$ is defined over a number field. So the eigenvalues λ_i of the residues of the Gauß-Manin connections lie in $\bar{\mathbb{Q}}$. On the other hand, the Gauß-Manin local system is defined over $\mathbb{Z}$, as this is the variation of the Betti cohomology of g, so the eigenvalues of the monodromy at infinity lie in $\bar{\mathbb{Q}}$. By [Del70, Corollaire 5.6, p. 96] $\mu_i = \exp(2\pi\sqrt{-1}\lambda_i)$. We conclude by Gelfond's theorem that $\lambda_i \in \mathbb{Q}$.

A second reason is as follows. Let S be an affine scheme of finite type over $\mathbb{Z}$ with $\mathcal{O}(S) \subset \mathbb{C}$, such that $X \hookrightarrow \bar{X}$ and a given complex point $x \in X$ have a model $X_S \hookrightarrow \bar{X}_S$ as a relative good compactification and x_S as an S-point of X_S. We assume moreover that the orders of the eigenvalues of the T_i are invertible on S. For any closed point $s \in |S|$ of residue field $\mathbb{F}_q$ of characteristic $p > 0$, with a $\bar{\mathbb{F}}_p$-point $\bar{s}$ above it, we denote by

$$sp_{\mathbb{C},\bar{s}} : \pi_1(X_{\mathbb{C}}, x_{\mathbb{C}}) \to \pi_1^t(X_{\bar{s}}, x_{\bar{s}})$$

the continuous surjective specialization homomorphism to the tame fundamental group [SGA1, Exposé XIII 2.10, Corollaire 2.12] and Sect. 4.3. Precomposing with the profinite completion homomorphism

$$\pi_1(X(\mathbb{C}), x(\mathbb{C})) \to \pi_1(X_{\mathbb{C}}, x_{\mathbb{C}})$$

yields

$$sp^{\mathrm{top}}_{\mathbb{C},\bar{s}} : \pi_1(X(\mathbb{C}), x(\mathbb{C})) \to \pi_1^t(X_{\bar{s}}, x_{\bar{s}})$$

which is compatible with the local fundamental groups, see [Del73, Section 1.1.10]. This enables one to transpose the quasi-unipotent monodromy condition to $X_{\bar{s}}$.

Finally we understand well how the Galois group Γ of $F = \mathrm{Frac}(\mathcal{O}(S))$ acts on the image $T_i^{\text{ét}}$ of the T_i in $\pi_1(X_{\mathbb{C}}, x_{\mathbb{C}})$, see [SGA7.2, XIV.1.1.10], [EK23, Lemma 2.1].

Lemma 6.1 *For each $1 \le i \le s$ the action of $\gamma \in \Gamma$ on π maps $T_i^{\text{ét}}$ to $(T_i^{\text{ét}})^{\chi(\gamma)}$, where $\chi : \Gamma \to \widehat{\mathbb{Z}}^\times$ is the cyclotomic character.*

6.3 Density Theorem for Quasi-unipotent Local Systems

Theorem 6.2 ([EK23], Theorem 1.3) *The set of $\rho \in Ch(\pi, r)^\square(\mathbb{C})$ with quasi-unipotent monodromy at infinity is Zariski dense in $Ch(\pi, r)^\square(\mathbb{C})$.*

Proof Let $Q \subset Ch(\pi, r)^\square(\pi)(\mathbb{C})$ be the Zariski closure of the set of quasi-unipotent representations. Assume $Q \neq Ch(\pi, r)^\square(\mathbb{C})$.

For each local monodromy at infinity $T_i \subset \pi$, choose $g_i \in T_i$. We have a morphism

$$\begin{array}{ccc} & & M = \mathbb{G}_m^s \\ & & \downarrow \varphi \\ Ch(\pi, r) & \xrightarrow{\psi} & N = \prod_{i=1}^s (\mathbb{A}^{r-1} \times \mathbb{G}_m) \end{array}$$

of affine schemes of finite type over $\mathbb{C}$ defined for each $i = 1, \ldots, s$ by the coefficients $(\sigma_1(\rho(g_i)), \ldots, \sigma_r(\rho(g_i))) \in N$ of the characteristic polynomials

$$\det(T \cdot \mathbb{I}_r - \rho(g_i)) = T^r - \sigma_1(\rho(g_i))T^{r-1} + \ldots + (-1)^r \sigma_r(\rho(g_i))$$

of a representation $\rho\colon \pi \to GL_r(\mathbb{C})$. (The morphism $\psi^\square$ factors through $Ch(\pi, r)$, but we do not use $Ch(\pi, r)$ in this chapter). There is a scheme B of finite type over

$\mathbb{Z}$ with factorization $\mathbb{Z} \to \mathcal{O}(B) \to \mathbb{C}$ over which the diagram $(\psi^{\square}, \varphi)$ and the inclusion $Q \hookrightarrow Ch(\pi, r)^{\square}$ are defined. We write $Q_B \hookrightarrow Ch(\pi, r)^{\square}_B$ and

$$\begin{array}{ccc} & & M_B = \mathbb{G}^s_{m,B} \\ & & \downarrow \varphi_B \\ Ch(\pi, r)^{\square}_B & \xrightarrow{\psi^{\square}_B} & N_B = \prod_{i=1}^s (\mathbb{A}^{r-1}_B \times \mathbb{G}_{m,B}) \end{array}$$

Using the section $\Gamma \to \pi_1(X_F, x_{\mathbb{C}})$ given by x_F, the Galois group Γ acts by conjugacy on $\pi_1(X_{\mathbb{C}}, x_{\mathbb{C}})$, thus on the set of closed points $|Ch(\pi, r)^{\square}|$. Since a closed point $z \in |Ch(\pi, r)^{\square}|$ has finite monodromy, its stabilizer $\Gamma_z \subset \Gamma$ is an open subgroup. It acts on the completion $(\widehat{Ch(\pi, r)^{\square}})_z$ at z. Said differently, the action $\Gamma \to \mathrm{Aut}(|Ch(\pi, r)^{\square}|)$ is continuous. Furthermore, Lemma 6.1 enables one to extend the action of Γ on the diagram $(\psi^{\square}_B, \varphi_B)$ in a compatible way with the action on $|Ch(\pi, r)^{\square}|$.

Set T_B to be the reduced Zariski closure of $\mathrm{Im}(\psi_B)$ and $S_B = \varphi_B^{-1}(T_B)$. As $Ch(\pi, r)^{\square}(\mathbb{C}) \setminus Q \neq \emptyset$, $Ch(\pi, r)^{\square}_B \setminus Q_B$ dominates B. So in particular, T_B and thus S_B dominate B as well. By generic smoothness, the smooth locus S^{sm}_B over B dominates B. By generic flatness for ψ_B restricted to $Ch(\pi, r)^{\square}_B \setminus Q_B$, its image meets $\varphi_B(S^{\mathrm{sm}}_B)$ (recall φ_B is finite). So there is a closed point $z \in |Ch(\pi, r)^{\square}_B \setminus Q_B|$ such that

- $\psi^{\square}$ is flat at z,
- $y = \psi^{\square}_B(z) \in \varphi_B(S^{\mathrm{sm}}_B)$.

We also fix a closed point $x \in S^{\mathrm{sm}}_B \cap \varphi_B^{-1}(y)$. Let Γ' be the intersection of stabilizers $\Gamma_x \cap \Gamma_z$, which is thus open in Γ, and let $b \in B$ be the image of the points x, y, z. So the closed subscheme $(\widehat{S_B})_x \hookrightarrow (\widehat{M_B})_x$ is Γ'-stable. We abuse notation and set $\Gamma' = \Gamma$.

We now take $s \in |S|$ a non-ramified closed point with residue field $\mathbb{F}_q$ of characteristic $p > 0$ different from the residual characteristic ℓ of x. Then the Frobenius $Frob_s$ lies in Γ and by Lemma 6.1 it acts on M_B and $(\widehat{M_B})_x$ by multiplication with q. By (as we are over $(\widehat{B})_b$, a variant of) de Jong's conjecture [dJ01, Conjecture 1.1] solved by Böckle-Khare [BK06] in specific cases and by Gaitsgory [Gai07] in general for $\ell \geq 3$, there are $Frob_s$-invariant points in $(\widehat{S_B})_x$, flat over $(\widehat{B})_b$. So there are points for which the coordinates in the group scheme M_B are $(q-1)$ roots of 1 which are flat over $(\widehat{B})_b$. By flatness of ψ_B restricted to $(\widehat{Ch(\pi, r)^{\square}_B \setminus Q_B})_z$, there is a point in $(\widehat{Ch(\pi, r)^{\square}_B} \setminus Q_B)_z$, flat over $(\widehat{B})_b$. This yields a complex topological local system outside of Q with eigenvalues of the T_i being $(q-1)$ roots of 1, a contradiction. See [EK23] for more details.

□

6.4 Remarks

(1) When I lectured on zoom in December 2020, at the pic of corona, on our density Theorem 6.2 with Moritz Kerz, Ben Bakker and Yohan Brunebarbe listened to the talk. They later on explained to us that they had a Hodge theoretical proof of the result. This would be nice, as it would add one stone to the line of similarities between complex and arithmetic methods.

(2) It would also be nice to single out subloci of the one consisting of complex local systems with quasi-unipotent monodromies at infinity for which density is preserved. One element of answer is provided by [dJE22, Section 3], see Sect. 6.7, in which we define the notion of weakly arithmetic local systems and prove their density. Weakly arithmetic local systems have in particular quasi-unipotent monodromies at infinity.

(3) Of course, we should remark that if X was projective to start with, our Theorem 6.2 would be void. Still if we think of varieties defined over number fields and Belyi's theorem, $X = \mathbb{P}^1 \setminus \{0, 1, \infty\}$ is the key scheme for many problems and so clearly the monodromies at infinity span the whole fundamental group. In addition if we think of analogies with the number theory case, where we look at Galois representations, we have ramification at bad primes. Unlike what was hoped for in [EK20, Question 9.1 (1)] and [EK23, Conjecture 1.1], the sublocus of arithmetic points is not dense, thanks to the work of Biswas-Gupta-Mj-Whang [BGMW22] in rank 2 and Landesman-Litt [LL22a], [LL22b] in any rank, see also [Lam22]. Some points of their construction is discussed in Sect. 6.6. Recall that given a smooth quasi-projective variety X and a semi-simple representation $\bar{\rho} : \pi_1(X, x) \to GL_r(\mathbb{F}_{\ell^a})$ for some $a \in \mathbb{N}_{>0}$, Mazur (and then Chenevier if $\bar{\rho}$ is not absolutely irreducible) associate to those data a deformation space of $\bar{\rho}$. This is a formal scheme over $W(\mathbb{F}_{\ell^a})$ with the property that its $\bar{\mathbb{Z}}_\ell$-points parametrize isomorphism classes of $\bar{\mathbb{Q}}_\ell$-local systems the semi-simplified residual representation of which is isomorphic to $\bar{\rho}$ (see [Maz89], [Che14] and [EK22, Proposition 4.3] specifically for the description of the local systems.) It would truly be bad if the sublocus of arithmetic points was not dense. In rank one it is, see [EK21, Theorem 1.3]. Over $\mathbb{C}$, as already mentioned in the previous paragraph, weakly arithmetic complex local systems are dense.

6.5 Some Other Dense or Not Dense Loci

We use the notation of Theorem 6.2 and fix some finite order conjugacy classes T_i of monodromies at infinity. Let $\mathcal{T} \subset M_B(X, r)(\mathbb{C})$ be the set of complex points with finite monodromy and $\mathcal{T}(T_i) \subset M_B(X, r, T_i)(\mathbb{C})$ be its intersection with $M_B(X, r, T_i)$. If $\mathcal{T}(T_i)$ is not empty, all the T_i must have finite order. We denote by $\bar{\mathcal{T}}$, resp. $\overline{\mathcal{T}(T_i)}$ the Zariski closure of $\mathcal{T}$ in $M_B(X, r)$, resp. $\mathcal{T}(T_i)$ in $M_B(X, r, T_i)$

and similarly by $\bar{\mathcal{T}}^{an}$ resp. $\overline{\mathcal{T}(T_i)}^{an}$ the analytic closures. For any smooth complex variety Y, we denote by $j : Y \hookrightarrow \bar{Y}$ a smooth projective compactification such that $\bar{Y} \setminus Y$ is a normal crossings divisor.

Proposition 6.3

(1) *If all the T_i have finite order, there is a finite Galois étale cover $h_1 : Y_1 \to X$ such that for any $\mathbb{L} \in M_B(X, r, T_i)(\mathbb{C})$, $h^*\mathbb{L}$ extends to a local system on $\bar{Y}_1$.*
(2) *There is a finite Galois étale cover $h_2 : Y_2 \to X$ such that*

  (a) *for any $\mathbb{L} \in \mathcal{T}$, $h^*\mathbb{L}$ is a sum of r torsion rank 1 local systems;*
  (b) *for any $\mathbb{L} \in \bar{\mathcal{T}}$, $h^*\mathbb{L}$ is a sum of r rank 1 local systems.*

(3) *If all the T_i have finite order, there is a finite Galois étale cover $h_3 : Y_3 \to X$ such that*

  (a) *for any $\mathbb{L} \in \overline{\mathcal{T}(T_i)}$, $h^*\mathbb{L}$ extends to a sum of r rank 1 local systems on $\bar{Y}_3$;*
  (b) *for any $\mathbb{L} \in \overline{\mathcal{T}(T_i)}^{an}$, $h^*\mathbb{L}$ extends to a sum of r rank 1 unitary local systems on $\bar{Y}_3$.*

Proof We prove (1). We consider the affine closed subschemes

$$\begin{array}{ccc} M_B(X, r, T_i)^{\square} & \longrightarrow & M_B(X, r)^{\square} \\ \downarrow q & & \downarrow q \\ M_B(X, r, T_i) & \longrightarrow & M_B(X, r) \end{array}$$

As $M_B(X, r, T_i)^{\square}$ is a fine moduli space, we have the universal representation

$$\rho_{univ} : \pi_1(X(\mathbb{C}), x(\mathbb{C})) \to GL_r(\mathcal{O}(M_B(X, r)^{\square})).$$

As $\pi_1(X(\mathbb{C}), x(\mathbb{C}))$ is finitely generated, Selberg's theorem (as used in [And04, Sections 8]) implies that there is a finite étale Galois cover $h_1 : Y_1 \to X$ such that $\rho_{univ}(\pi_1(Y_1(\mathbb{C}), y_1(\mathbb{C})))$ is torsion-free. Here $y_1 \in Y_1$ lies above x. As the T_i were assumed to have finite order, this implies that $\rho_{univ}|_{(\pi_1(Y_1(\mathbb{C}), y_1(\mathbb{C})))}$ factors through $\pi_1(\bar{Y}_1(\mathbb{C}), y_1(\mathbb{C}))$. Any complex point $\rho \in M_B(X, r, T_i)^{\square}(\mathbb{C})$ is defined as the local system to $\iota \circ \rho_{univ}$, where $\iota : \mathcal{O}(M_B(X, r, T_i)^{\square}) \to \mathbb{C}$ is a complex point. We conclude that all complex points of $M_B(X, r, T_i)$, restricted to Y_1, extends to $\bar{Y}_1$, that is they have no ramification. This proves (1).

We prove (2)(a). We use Jordan's theorem as in [And04, Section 10]: there is a finite étale Galois cover $h_2 : Y_2 \to X$ such that for all $\mathbb{L} \in \mathcal{T}$, $h^*\mathbb{L}$ is a sum of r torsion rank 1 local systems. This proves (2)(a).

We prove (2)(b). The morphism $h_2^* : M_B(X, r) \to M_B(Y, r)$ is finite onto its image, thus $h^*(\bar{\mathcal{T}})$ is the Zariski closure of $h^*(\mathcal{T})$ in $M_B(Y, r)$. On the other hand, the locus in $M_B(Y, r)$ consisting sums of r rank 1 local systems is Zariski closed, see Lemma 6.5. This proves (2)(b).

We prove (3)(a). See also [EL13, Theorem 5.1] for an analogous statement for vector bundles.

Taking now $h_3 : Y_3 \to X$ finite étale Galois dominating both h_1 and h_2, (1) holds for h_3. The extension to $\bar{Y}_3$ defines the factorization

$$h_3^* : M_B(X, r, T_i) \xrightarrow{c^*} M_B(\bar{Y}_3, r) \xrightarrow{j^*} M_B(Y_3, r)$$

where h_3^* is finite and its image is Zariski closed. So c^* is finite and its image is Zariski closed. As j^* is furthermore Zariski closed, the Zariski closure of $h_3^*(\mathcal{T}(T_i))$ in $M_B(Y_3, r)$ is the image by j^* of the Zariski closure of $c^*(\mathcal{T}(T_i))$ in $M_B(\bar{Y}_3, r)$ which itself is equal to $c^*(\overline{\mathcal{T}(T_i)})$. We apply Lemma 6.5 to conclude that $c^*(\mathcal{T}(T_i))$ consists of local systems which split as a sum $\oplus_1^r \mathcal{L}_i$ of r rank 1 local systems. This proves (3)(a).

We prove (3)(b). We have to see that if $\mathbb{L} \in \overline{\mathcal{T}(T_i)}^{an}$ then the $\mathcal{L}_i$ are unitary. The analytic closure of torsion rank 1 local systems on $M_B(\bar{Y}_3, 1)$ is the space $U(\bar{Y}_3)$ of unitary rank 1 connections, which is compact. Thus via the algebraic morphism $M_B(\bar{Y}_3, 1)^r \to M_B(\bar{Y}_3, r)(\hookrightarrow M_B(Y_3, r))$ the image of $U(\bar{Y}_3)^r \to M_B(\bar{Y}_3, r)(\hookrightarrow M_B(Y_3, r))$ is analytically closed. Thus $c^*(\overline{\mathcal{T}(T_i)}^{an})$ lies in this image. This proves (3)(b).

□

Remark 6.4 In Proposition 6.3 we cannot replace in 3) $\overline{\mathcal{T}(T_i)}$ by $\bar{\mathcal{T}}$ and similarly for the analytic version. Assume $X = \mathbb{G}_m$ then $\pi_1(X(\mathbb{C}), x(\mathbb{C})) = \mathbb{Z}$ and $M_B(X, 1) = \mathbb{G}_m$, where for $\mu \in \mathbb{G}_m(\mathbb{C}) = \mathbb{C}^\times$, the associated local system $\mathbb{L}$ is defined by $\mathbb{Z} \to \mathbb{C}^\times$, $1 \mapsto \mu$. Then $\mathcal{T} = \mu_\infty \subset \mathbb{C}^\times$, the subgroup of roots of unity, its analytic closure is $S^1 \subset \mathbb{C}^\times$, the circle, its Zariski closure is the whole of $\mathbb{C}^\times$. However, $h : Y \to \mathbb{G}_m$ has the property that $\bar{Y}$ is isomorphic to $\mathbb{P}^1$, thus is simply connected and non-torsion rank 1 local systems cannot be trivialized by a finite étale cover. See also [And04, Remarque 7.2.3] for the same example used in a similar spirit.

The following lemma ought to be well known, we write a short proof as we could not find it in the literature.

Lemma 6.5 *For any finitely generated group π, and natural number $r \geq 1$, the locus of complex points $\mathbb{L}$ in $Ch(\pi, r)$ such that $\mathbb{L}$ is a sum of rank 1 representations up to isomorphism is closed.*

Proof The structure morphism $q : Ch(\pi, r)^\square \to Ch(\pi, r)$ endows $Ch(\pi, r)$ with the quotient topology, so Lemma 6.5 is equivalent to saying that the locus of complex points ρ in $Ch(\pi, r)^\square$ which stabilize a complete flag, i.e. the associated graded splits $\mathbb{C}^r$ as a sum of rank 1 vector subspaces, is Zariski closed. Let $(1 = \gamma_1, \gamma_2, \ldots, \gamma_s)$ be generators of π. Let Fl be the variety of complete flags on $\mathbb{C}^r$. Consider the algebraic map

$$Ch(\pi, r)^\square \to \mathrm{Hom}(Fl, (Fl)^s), \ (\rho \mapsto [Fl \ni F \mapsto (Fl)^s \ni (\gamma_1(F), \ldots, \gamma_s(F)]).$$

Let $\Delta \subset (Fl)^s$ be the diagonal. Then $\mathrm{Hom}(Fl, \Delta) \subset \mathrm{Hom}(Fl, (Fl)^s)$ is a closed embedding, so its inverse image in $Ch(\pi, r)^{\square}$ is closed as well.

□

Proposition 6.3 (2)(b) is a convenient way to see that some $\mathbb{L}$ cannot be in $\bar{\mathcal{T}}$, see Example 6.6 and Corollary 6.8.

Example 6.6 Let $\rho : \pi_1(X(\mathbb{C}), x(\mathbb{C})) \to GL_r(\mathbb{C})$ be a representation with the property that $\rho(\pi_1(X(\mathbb{C}), x(\mathbb{C})))$ is Zariski dense in $G \subset GL_r(\mathbb{C})$, where $G = GL_{r'}(\mathbb{C})$ or $G = SL_{r'}(\mathbb{C})$ for some r' with $2 \leq r' \leq r$. Then ρ is semi-simple and its moduli point does not lie in $\bar{\mathcal{T}}$.

6.6 Arithmetic Local Systems Are Not Dense: The Work of Biswas-Gupta-Mj-Whang and Landesman-Litt

We sketch some aspects of the construction referred to in Remark 6.4 (2) by Aaron Landesman and Daniel Litt, which generalizes in higher rank Theorem A and Lemma 3.2 of [BGMW22]. While the latter uses the specificity of SL_2 representations, which have finite monodromy if and only if they do on specific loops, the former uses Hodge theory. We are grateful to the two authors for their swift, precise and friendly answers to our questions. We just develop some points of the arguments in a simplified geometric situation and refer to their original articles [LL22a, LL22b] for the complete statements and proofs.

6.6.1 *Statement*

Let $f : X \to S$ be a smooth family of genus ≥ 2 curves together with a dominant morphism $h : S \to M_g$, where M_g is the coarse moduli space of genus g curves. We shall assume that f has a section $S \to X$ which is irrelevant for the statements. Let $s \in S$ be a complex generic point. We denote by X_s the fibre of f above s. For $x \in X_s$ a geometric point, for example the one on the section of f, the homotopy sequence

$$1 \to \pi_1(X_s(\mathbb{C}), x(\mathbb{C})) \to \pi_1(X(\mathbb{C}), x(\mathbb{C})) \to \pi_1(S(\mathbb{C}), s(\mathbb{C})) \to 1$$

is exact. As the image of $\pi_1(X_s(\mathbb{C}), x(\mathbb{C})) \to \pi_1(X(\mathbb{C}), x(\mathbb{C}))$ is normal, the restriction $\mathbb{L}|_{X_s}$ to X_s of any semi-simple complex local system $\mathbb{L}$ on X is semi-simple.

Theorem 6.7 ([LL22b], Theorem 1.2.1) *If a semi-simple local system $\mathbb{L}$ on X has rank $r < \sqrt{g+1}$, then $\mathbb{L}|_{X_s}$ has finite monodromy.*

If $r = 1$, so in particular if $g = 1$, Theorem 6.7 is due to [BGMW22, Lemma 3.2] by analyzing the effect of the Dehn twists on the images of the generators of $H_1(X_x, \mathbb{Z})$ under the character representation. If $r = 2$ this is due to [BGMW22, Theorem A], by analyzing the monodromy on specific loops. Also the theorem is valid if $\mathbb{L}$ is not semi-simple. We won't discuss this last step.

Corollary 6.8 ([LL22a], Corollary 1.2.10, Lemma 7.5.1) *If $g \geq 2$ and $2 \leq r < \sqrt{g+1}$, the image of the restriction morphism $M_B(X, r) \to M(X_s, r)$ is not dense.*

Proof of Corollary 6.8 We have to show that the moduli points corresponding to local systems with finite monodromy are not Zariski dense. Let $a_1, b_1, \ldots, a_g, b_g,\ g \geq 2$ be generators of $\pi_1(X_s(\mathbb{C}), x(\mathbb{C}))$ with the single relation $\prod_{i=1}^{g} [a_i, b_i] = 1$. Define $\rho : \pi_1(X_s(\mathbb{C}), x(\mathbb{C})) \to GL_r(\mathbb{C})$ by sending a_1, a_2 to the block matrices

$$\begin{pmatrix} 1\,1 & 0 \\ 0\,1 & 0 \\ 0\,0 & \mathbb{I}_{r-2} \end{pmatrix} \quad \begin{pmatrix} 1\,0 & 0 \\ 1\,1 & 0 \\ 0\,0 & \mathbb{I}_{r-2} \end{pmatrix}$$

and all the other generators a_j for $j \geq 3$ and b_i for $i \geq 1$ to $\mathbb{I}_r$, where $\mathbb{I}_n$ is the unit square matrix of size $n \times n$. The monodromy group is dense in $SL_2(\mathbb{C}) \subset GL_r(\mathbb{C})$. We apply Example 6.6. □

Remark 6.9 As $M_B(\pi_1(X(\mathbb{C}), x(\mathbb{C})), r)^{\square} \to M(X, r)$ is surjective, we could equally say in Corollary 6.8 that the image of $M_B(\pi_1(X(\mathbb{C}), x(\mathbb{C})), r)^{\square} \to M(X_s, r)$ is not dense. Said in words, a semi-simple local system on X_s which lifts to X lifts as a semi-simple local system, and those are not dense for r small.

We keep the same notation as above. The morphism $f : X \to S$ descends to $f_\circ : X_\circ \to S_\circ$ over $F_\circ$ where $F_\circ/\mathbb{Q}$ is a field of finite type over $\mathbb{Q}$. Thus $F_\circ(S_\circ) \subset \kappa(x) \cong \mathbb{C}$, the residue field of x. Recall that an ℓ-adic local system $\mathbb{L}_\ell$ on X_s is arithmetic if it descends to X_F, where $F_\circ(S_\circ) \subset F(\subset \kappa(x))$ is a finite type extension. Geometrically, it means that there is a dominant morphism $T_1 \to S_\circ \otimes_{F_\circ} F_1$ defined over F_1, a finite extension of $F_\circ$ such that $F = F_1(T_1)$, and such that $\mathbb{L}_\ell$ is defined over the fibre of the pull-back morphism $f_\circ \otimes_{S_\circ} T_1 =: f_1 : X_\circ \otimes_{S_\circ} T_1 =: X_1 \to T_1$ over the generic point $\mathrm{Spec}(F)$. As the notion of arithmeticity does not depend on the choice of such an F, see [dJE22, Remark 3.2], we'll assume henceforth that f_1 admits a section $T_1 \to X_1$, and then we abuse notation setting $f = f_1$, so also $F = F_\circ(S)$. In particular, this guarantees that the homotopy sequence of topological fundamental groups

$$1 \to \pi_1(X_s(\mathbb{C}), x(\mathbb{C})) \to \pi_1(X(\mathbb{C}), x(\mathbb{C})) \to \pi_1(S(\mathbb{C}), s(\mathbb{C})) \to 1$$

is exact, and by [And74, Proposition 3], if $g \geq 2$ which implies that the center of $\pi_1(X_s(\mathbb{C}), x(\mathbb{C}))$ is trivial, that the induced homotopy sequence of étale fundamental groups

$$(\star)\ 1 \to \pi_1(X_s, x) \to \pi_1(X, x) \to \pi_1(S, s) \to 1$$

is exact as well.

Corollary 6.10 ([LL22b], Theorem 8.1.2) *If $g \geq 2$ and $2 \leq r < \sqrt{g+1}$, the arithmetic local systems in $M_B(X_s, r)(\bar{\mathbb{Q}}_\ell)$ are not Zariski dense.*

Proof The action of $\mathrm{Gal}(\bar{F}/F)$ on ℓ-adic local systems comes from the homotopy exact sequence

$$1 \to \pi_1(X_{\bar{x}}, x) \to \pi_1(X_F, x) \to \mathrm{Gal}(\bar{F}/F) \to 1$$

and the action of $\mathrm{Gal}(\bar{F}/F)$ on $\pi_1(X_{\bar{x}}, x)$ via outer automorphisms (which can be lifted to automorphisms thanks to the section). An ℓ-adic arithmetic local system $\mathbb{L}_{\ell,\mathbb{C}}$ descending to $\mathbb{L}_{\ell,F'}$, for $F \subset F'(\subset \bar{F})$ a finite Galois extension, is a fixpoint under this action restricted to $\mathrm{Gal}(\bar{F}/F')$. We may assume F is equal to F', and S', the normalization of S in F', is equal to S. We restrict this action to $\mathrm{Gal}(\overline{\mathbb{C}(S)}/\mathbb{C}(S))$ via the base change morphism $\mathrm{Gal}(\overline{\mathbb{C}(S)}/\mathbb{C}(S)) \to \mathrm{Gal}(\bar{F}/F)$ from $F = F_\circ(S)$ to $\mathbb{C}(S)$. By left exactness of $(\star)$, the action factors through $\pi_1(S, s)$. Thus in fact $\mathbb{L}_{F\otimes_{F_\circ}\mathbb{C}}$ descends to an ℓ adic local system on X. Viewed as a topological local system with $\bar{\mathbb{Q}}_\ell$-coefficients, its restriction to X_s has finite monodromy by choosing randomly a field isomorphism between $\bar{\mathbb{Q}}_\ell$ and $\mathbb{C}$ and applying Theorem 6.7. We apply Corollary 6.8. This finishes the proof. □

6.6.2 Rigidity

The notation is as in Sect. 6.6.1. We had already discussed that an irreducible local system $\mathbb{L}$ has finite determinant. Moreover, by [LLSS20, Lemma 2.1.1], [AS16, Proposition 2.4] and Lefschetz theory which implies that the whole monodromy is recognized on good curves, it has quasi-unipotent monodromies at infinity.

Recall that an irreducible complex local system $\mathbb{L}$ with finite determinant and quasi-unipotent monodromies at infinity on a smooth variety X is called *strongly cohomologically rigid* if

$$H^1(X, \mathcal{E}nd^0(\mathbb{L})) = 0.$$

Here the upper index 0 indicates the trace free endomorphisms. This strong rigidity notation was introduced and discussed in [EG21, Definition A.1.b)], as it is verified for any irreducible local system on a Shimura variety of real rank ≥ 2. It implies in particular that $\mathbb{L}$ is *cohomologically rigid* as the map

$$IC^1(X, \mathcal{E}nd^0(\mathbb{L})) \to H^1(X, \mathcal{E}nd^0(\mathbb{L}))$$

from intersection cohomology to cohomology is injective. Recall the integrality theorem [EG18, Theorem 1.1], see the report on it in Sect. 7.5.

Theorem 6.11 *If $\mathbb{L}$ is cohomologically rigid, then it is integral, so in particular so is its restriction $\mathbb{L}|_{X_s}$.*

6.6.3 Proof of Theorem 6.7

To show finiteness of the monodromy of $\mathbb{L}|_{X_s}$, we thus only need two steps:

(I) For $r < \sqrt{g+1}$, $\mathbb{L}|_{X_s}$ is a direct sum of irreducible unitary local systems U_i.

If so, as by the theorem of Krull-Schmidt this decomposition is unique, the action of $\pi_1(S(\mathbb{C}), s(\mathbb{C}))$ on the set $\{U_i\}_i$ is finite, so after a finite étale base change $S' \to S$, U_i lifts to $\mathbb{U}_i$ to $X' := S' \times_S X$. So for the problem, we may assume $S' \to S$ is the identity.

(II) $\mathbb{U}_i$ is strongly cohomologically rigid.

If so, then applying Theorem 6.11 to the conjugates $\mathbb{L}^\sigma$ on the coefficients of the monodromies of $\mathbb{L}$, the local system $(\oplus_\sigma \mathbb{L}^\sigma)|_{X_s}$ is then unitary and has by Theorem 6.11 monodromy defined over $\mathbb{Z}$, thus has finite monodromy.

Ad (II): once we have the $\mathbb{U}_i$, as U_i is irreducible, $f_*\mathcal{E}nd^0(U_i) = 0$ thus the Leray spectral sequence yields

$$H^1(X, \mathcal{E}nd^0(\mathbb{U}_i)) = H^0(S, R^1 f_*\mathcal{E}nd^0(\mathbb{U}_i))$$

and (II) is equivalent to

$$H^0(S, R^1 f_*\mathcal{E}nd^0(\mathbb{U}_i)) = 0.$$

This vanishing is going to be handled in I)c) below.

Ad (I): starting with $\mathbb{L}$, as the subgroup $\pi(X_s(\mathbb{C}), x(\mathbb{C})) \subset \pi_1(X(\mathbb{C}), x(\mathbb{C}))$ is normal, the restriction $\mathbb{L}|_{X_s}$ is semi-simple. So one wants to prove that the summands are all unitary. The way the authors proceed is as follows.

(I)(a) Show (I) is true if $\mathbb{L}$ is a polarized complex variation of Hodge structures. This is the content of [LL22a, Theorem 1.2.12] for which the bound is slightly better, one needs only $r < 2\sqrt{g+1}$.

(I)(b) If $\mathbb{L}$ is not from the beginning a polarized complex variation of Hodge structures, we know by Simpson-Mochizuki [Moc06, Theorem 10.5] that the connected component of $M_B(X, r)$ which contains the moduli point to $\mathbb{L}$

contains also the moduli point of one such polarized complex variation of Hodge structure which we denote by $\mathbb{M}$.

(I)(c) Show that $\mathbb{L}|_{X_s} = \mathbb{M}|_{X_s}$. For this, one needs in the proof the sharper bound $r < \sqrt{g+1}$.

Let (E, ∇) be the algebraic flat connection associated to $\mathbb{L}$. The $\mathbb{L}|_{X_s}$ is unitary if and only if $E|_{X_s}$ is (slope) semi-stable. The proof of (I)(a) computes semi-stability in low rank.

Ad (I)(c): recall that the tangent map of the restriction $M_B(X, r) \to M_B(X_s, r)$ at the point $\mathbb{M}$ is identified with the restriction map

$$H^1(X, \mathcal{E}nd^0(\mathbb{M})) \to H^1(X_s, \mathcal{E}nd^0(\mathbb{M})).$$

This map factors through

$$H^0(S, R^1 f_* \mathcal{E}nd^0(\mathbb{M})) \subset H^1(X_s, \mathcal{E}nd^0(\mathbb{M})).$$

So the first step for proving (I)(c) is to show the vanishing of $H^0(S, R^1 f_* \mathcal{E}nd^0(\mathbb{M}))$, which in particular implies the vanishing needed for (II) and so settles (II). This is performed using Deligne's fix part theorem (see [LL22b, Theorem 1.7.1]). If both $\mathbb{M}$ and $\mathbb{M}|_{X_s}$ were irreducible, the formality theorem of Goldman-Millson [GM88, Proposition 4.4] applied to X_s and to X (by a wishful generalization to the non-proper case) would enable to conclude that the homomorphism of completed local rings $\hat{\mathcal{O}}_{M_B(X_s,r),\mathbb{M}|_{X_s}} \to \hat{\mathcal{O}}_{M_B(X,r),\mathbb{M}}$ factors through $\hat{\mathcal{O}}_{M_B(X_s,r),\mathbb{M}|_{X_s}} \xrightarrow{\mathbb{M}|_{X_s}} \mathbb{C}$, the closed point corresponding to $\mathbb{M}|_{X_s}$, and consequently the restriction map $M_B(X, r) \to M_B(X_s, r)$ contracts the connected component containing $\mathbb{M}$ to its restriction $\mathbb{M}|_{X_s}$.

The proof proceeds by showing the vanishing

$$H^0(S, R^1 f_* \mathcal{E}nd^0(\mathbb{M}_A)) = 0$$

for any extension of $\mathbb{M}$ to a local system $\mathbb{M}_A$ with coefficients in any local Artinian ring A and for rank $\mathcal{E}nd^0(\mathbb{M}) < g$, that is $r < \sqrt{g+1}$. This is an essential point of the proof and is performed using the Kodaira-Spencer class combined with Deligne's fix part theorem (see [LL22b, Theorem 6.2.1]). This implies that the tangent map of $M_B(X, r) \to M_B(X_s, r)$ at the A-point $\mathbb{M}_A$ dies. As this is true for all local Artinian rings A, this shows that $\hat{\mathcal{O}}_{M_B(X_s,r),\mathbb{M}|_{X_s}} \to \hat{\mathcal{O}}_{M_B(X_s,r),\mathbb{M}|_{X_s}}$ factors through $\hat{\mathcal{O}}_{M_B(X_s,r),\mathbb{M}|_{X_s}} \xrightarrow{\mathbb{M}|_{X_s}} \mathbb{C}$ (see [LL22b, Proposition 3.2.1]).

6.7 Weakly Arithmetic Complex Local Systems

In this section we report on the notion of weakly arithmetic local systems introduced with Johan de Jong in [dJE22, Section 3] which is weaker than the notion of arithmetic local systems. With this weaker notion of arithmeticity, we can prove density in the Betti moduli space. We refer to *loc. cit.* for complete statements and proofs.

6.7.1 *Definitions*

Let X be a smooth complex quasi-projective variety. We fix a projective compactification $X \hookrightarrow \bar{X}$ such that $\bar{X} \setminus X$ is a strict normal crossings divisor. The compactification $X \hookrightarrow \bar{X}$ is defined over a field of finite type $F \subset \mathbb{C}$, and

$$(X \hookrightarrow \bar{X}) = (X_F \hookrightarrow \bar{X}_F) \otimes_F \mathbb{C}.$$

It defines the algebraic closure $F \subset \bar{F} \subset \mathbb{C}$ of F. Recall that the homotopy exact sequence

$$1 \to \pi_1(X_{\bar{F}}, x_{\bar{F}}) \to \pi_1(X_F, x_{\bar{F}}) \to \mathrm{Gal}(\bar{F}/F) \to 1$$

defines an action of $\mathrm{Gal}(\bar{F}/F)$ on ℓ-adic local systems coming from the outer action of $\mathrm{Gal}(\bar{F}/F)$ on $\pi_1(X_{\bar{F}}, x_{\bar{F}})$. An *arithmetic* ℓ-adic local system on $X_{\bar{F}}$ is one which has a finite orbit under this action. Equivalently it descends to $X_{F'}$ where $F \subset F' \subset \bar{F} \subset \mathbb{C}$ is a finite extension. Clearly for this definition we may replace F by a finite extension in $\bar{F}$ so assume that X has a rational point. Then the outer action becomes just an action. More generally, we may replace F by a finite type extension $F \subset F'$ in $\bar{F}$ as then the image of $\mathrm{Gal}(\bar{F}/F') \to \mathrm{Gal}(\bar{F}/F)$ is open so the definition does not depend on the F chosen (see e.g. [dJE22, Remark 3.2]).

We choose a model X_S of X over S a scheme of finite type over $\mathbb{Z}$, with field of fractions F, such that

$$(X_F \hookrightarrow \bar{X}_F, x_F) = (X_S \hookrightarrow \bar{X}_S, x_S) \otimes_S F$$

where $X_S \hookrightarrow \bar{X}_S$ is smooth relative compactification, $\bar{X}_S \setminus X_S$ is a relative normal crossings divisor. Then in fact an arithmetic local system $\mathbb{L}_\ell$ is *unramified almost everywhere* on S, that is there is a dense open subscheme $S^\circ \subset S$ such that $\mathbb{L}_\ell$, defined on X_F, descends to X_{S°, see e.g. [Pet22, Proposition 6.1] where the argument is performed for S the spectrum of a number ring, and see references therein.

We have the specialization homomorphism

$$sp^{\text{top}}_{\mathbb{C},\bar{F}} : \pi_1(X(\mathbb{C}), x(\mathbb{C})) \to \pi_1(X_{\bar{F}}, x_{\bar{F}})$$

which stems from the Riemann Existence Theorem equating $\pi_1(X_{\mathbb{C}}, x_{\mathbb{C}})$ with the profinite completion of $\pi_1(X(\mathbb{C}), x(\mathbb{C}))$, then the base change continuous isomorphism equating $\pi_1(X_{\mathbb{C}}, x_{\mathbb{C}})$ with the geometric étale fundamental group $\pi_1(X_{\bar{F}}, x_{\bar{F}})$, which we can then postcompose with the continuous embedding $\pi_1(X_{\bar{F}}, x_{\bar{F}}) \to \pi_1(X_F, x_F)$ from the the geometric étale fundamental group to the arithmetic one. In total this defines

$$sp^{\text{top}}_{\mathbb{C},F} : \pi_1(X(\mathbb{C}), x(\mathbb{C})) \to \pi_1(X_F, x_F).$$

For any closed point $s \in S$ with geometric point $\bar{s} \to S$ above it, we have a specialization homomorphism

$$sp_{\bar{F},\bar{s}} : \pi_1(X_{\bar{F}}, x_{\bar{F}}) \to \pi_1^t(X_{\bar{s}}, x_{\bar{s}})$$

which extends to

$$sp_{F,s} : \pi_1(X_F, x_{\bar{F}}) \to \pi_1^t(X_s, x_{\bar{s}}).$$

By precomposing with $sp^{\text{top}}_{\mathbb{C},\bar{F}}$, $sp^{\text{top}}_{\mathbb{C},F}$, we obtain

$$sp^{\text{top}}_{\mathbb{C},\bar{s}} : \pi_1(X(\mathbb{C}), x(\mathbb{C})) \to \pi_1^t(X_{\bar{s}}, x_{\bar{s}}),\ sp^{\text{top}}_{\mathbb{C},s} : \pi_1(X(\mathbb{C}), x(\mathbb{C})) \to \pi_1^t(X_s, x_{\bar{s}}).$$

where the upper script ^{t} indicates the tame fundamental group.

If $\tau : K_1 \to K_2$ is a homomorphism of fields, and $\mathbb{L}_1$ is a topological local system defined over K_1 by the homomorphism $\rho : \pi_1(X(\mathbb{C}), x(\mathbb{C})) \to GL_r(K_1)$, we denote by $\mathbb{L}^{\tau}_{K_1}$ the local system defined over K_2 obtained by post-composing ρ by the homomorphism $GL_r(K_1) \to GL_r(K_2)$ defined by τ.

Choosing quasi-unipotent conjugacy classes T_i in $GL_r(\mathbb{C})$ and a torsion rank 1 local system $\mathcal{L}$ on $X(\bar{\mathbb{C}})$, we recall that $M_B(X, r, \mathcal{L}, T_i)$ is defined over a number ring $\mathcal{O}_K \subset \mathbb{C}$. We say that a complex point $\mathbb{L}_{\mathbb{C}}$ is *weakly arithmetic* if made ℓ-adic for some ℓ, and descended to a tame ℓ-adic local system via $sp^{\text{top}}_{\mathbb{C},\bar{s}}$ for some closed point $s \in$ of large characteristic, it is on $X_{\bar{s}}$ arithmetic, that is it descends to $X_{s'}$ for $s' \to s$ a finite extension below $\bar{s}$. Formally, we write the definition as follows.

Definition 6.12 (See [dJE22], Definition 3.1) A point $\mathbb{L}_{\mathbb{C}} \in M_B(X, r, \mathcal{L}, T_i)(\mathbb{C})$ is said to be *weakly arithmetic* if there exist

(i) a prime number ℓ and a field isomorphism $\tau : \bar{\mathbb{Q}}_\ell \to \mathbb{C}$ (so τ defines an $\mathcal{O}_K$-algebra structure on $\bar{\mathbb{Q}}_\ell$ via $\mathcal{O}_K \subset \mathbb{C}$);
(ii) a finite type scheme S such that $(X \hookrightarrow \bar{X}, x, \mathcal{L}, T_i)$ has a model over S;

(iii) a closed point $s \in |S|$ of residual characteristic different from ℓ;
(iv) a tame arithmetic ℓ-adic local system $\mathbb{L}_{\ell,\bar{s}}$ on $X_{\bar{s}}$ with determinant $\mathcal{L}$ and monodromy at infinity in T_i;

such that $\mathbb{L}_{\mathbb{C}}$ and $\left((sp^{\text{top}}_{\mathbb{C},\bar{s}})^{-1}(\mathbb{L}_{\ell,\bar{s}})\right)^{\tau}$ are isomorphic.

Remark 6.13 The choice of τ prevents the set of complex points in $M_B(X, r, \mathcal{L}, T_i)$ which are the moduli points of weakly arithmetic local systems to be a priori countable, while the set of complex points in $M_B(X, r, \mathcal{L}, T_i)$ which are the moduli points of arithmetic local systems is really countable. So in the present state of understanding, this is a "poor" approximation.

Notations 6.14

(1) Fixing $(X, r, \mathcal{L}, T_i)$, we denote by

$$W(X, r, \mathcal{L}, T_i) \subset M_B(X, r, \mathcal{L}, T_i)(\mathbb{C})$$

the locus of weakly arithmetic local systems.

(2) Fixing $(X, r, \mathcal{L})$, we denote by

$$W(X, r, \mathcal{L}) = \cup_{\{T_i\}} W(X, r, \mathcal{L}, T_i) \subset M_B(X, r, \mathcal{L})(\mathbb{C})$$

the locus of all weakly arithmetic local systems of rank r and determinant $\mathcal{L}$.

6.7.2 Density

Theorem 6.15 (See [dJE22], Theorem 3.5)

(1) Fixing $(X, r, \mathcal{L}, T_i)$, $W(X, r, \mathcal{L}, T_i)$ is dense in $M_B(X, r, \mathcal{L}, T_i)(\mathbb{C})$.
(2) Fixing $(X, r, \mathcal{L})$, $W(X, r, \mathcal{L})$ is dense in $M_B(X, r, \mathcal{L})(\mathbb{C})$.

Proof Ad (1): Let $T_{\mathbb{C}}$ be the Zariski closure of $W(X, r, \mathcal{L}, T_i)$ in $M_B(X, r, \mathcal{L}, T_i)(\mathbb{C})$ and $T_{\mathcal{O}_K}$ be the Zariski closure of $T_{\mathbb{C}}$ in $M_B(X, r, \mathcal{L}, T_i)$. As $T_{\mathbb{C}}$ is invariant under the group $\text{Aut}_{\mathcal{O}_K}(\mathbb{C})$ of field automorphisms of $\mathbb{C}$ over $\mathcal{O}_K$, $T_{\mathcal{O}_K}$ is the Zariski closure of $W(X, r, \mathcal{L}, T_i)$ in $M_B(X, r, \mathcal{L}, T_i)$.

If $T_{\mathbb{C}}$ was not equal to $M_B(X, r, \mathcal{L}, T_i)(\mathbb{C})$, we could choose a closed point $z \in M_B(X, r, \mathcal{L}, T_i)$ of characteristic ≥ 3 which has irreducible monodromy. The quotient morphism

$$M_B(X, r, \mathcal{L}, T_i)^{\square} \to M_B(X, r, \mathcal{L}, T_i)$$

in a neighbourhood of z is smooth, so it makes sense to talk on the irreducibility of the monodromy. We assume in addition that it does not lie on $T_{\mathcal{O}_K}$, and that it is smooth for the projection $M_B(X, r, \mathcal{L}, T_i)(\mathbb{C})_{\text{red}} \to \text{Spec}(\mathcal{O}_K)$. We choose a closed point $s \in S^{\circ}$ of characteristic p larger than the cardinality of the monodromy

group of z, the order of $\mathcal{L}$ and of the eigenvalues of the T_i. So the torsion local system corresponding to z descends via $sp_{\bar{F},\bar{s}}$ to $\mathbb{L}_{z,\bar{s}}$, a local system on $X_{\bar{s}}$. Via the identification of Mazur's deformation ring $D_{z,\bar{s}}$ [Maz89, Proposition 1] with the formal completion of $M_B(X, r, \mathcal{L}, T_i)$ at z, simply comparing the moduli functors, see [dJE22, Proposition 2.2], we can apply Drinfeld's argument [Dri01, Lemma 2.8] to conclude: de Jong's conjecture [dJ01, Conjecture 2.3], proved by Böckle-Khare [BK06] in specific cases and by Gaitsgory [Gai07] for $\ell \geq 3$, implies that $\mathbb{L}_{z,\bar{s}}$ admits a lift as a $\bar{\mathbb{Z}}_\ell$-local system on $X_{\bar{s}}$ which is arithmetic. Then $(sp^{\text{top}}_{\mathbb{C},\bar{s}})^{-1}(\mathbb{L}_{\ell,\bar{s},z})$ does not lie in $T_{\mathcal{O}_K}(\bar{\mathbb{Q}}_\ell)$. By invariance, for any τ, $((sp^{\text{top}}_{\mathbb{C},\bar{s}})^{-1}(\mathbb{L}_{\ell,\bar{s},z}))^\tau$ does not lie on $T(\mathbb{C})$, a contradiction to the definition of weak arithmeticity.
Ad (2): By Theorem 6.2 [EK23, Theorem 1.3], for a torsion rank 1 local system $\mathcal{L}$ given,

$$\cup_{T_i} M_B(X, r, \mathcal{L}, T_i)$$

is dense in $M_B(X, r, \mathcal{L})(\mathbb{C})$. Combined with (1), this yields (2).

□

Remark 6.16 The notion of weakly arithmetic local systems in Definition 6.12 is forced upon us by Drinfeld's proof of Kashiwara's conjecture in complex geometry using de Jong's conjecture over finite fields in [Dri01]. Without really explicitly developing the concept, he proves density, and without explicitly mentioning it, he gives an arithmetic proof of Simpson's Hard-Lefschetz theorem for a semi-simple complex local system on a smooth complex projective complex variety X [Sim92, Lemma 2.6]. We do not detail this last point here, we refer to [EK21, Theorem 1.1] where we implement this strategy for the proof in rank $r = 1$ over $\bar{\mathbb{F}}_p$. There is no way we can generalize the $r = 1$ proof to higher rank. It uses strongly the shape of $M_B(X, 1)$, a commutative group scheme which is the extension of a finite group by a torus. In light of the proof of Theorem 6.15, one key point which is missing is the possibility to go from ℓ to ℓ', even in a non-canonical way (so without arithmeticity and companions, see Sect. 7). May be it is also related to the notion of p'-discrete generation (presentation) developed in Sect. 5.

Chapter 7
Lecture 7: Companions, Integrality of Cohomologically Rigid Local Systems and of the Betti Moduli Space

Abstract The notion of companions has been defined by Deligne (Publ Math IHÉS 52:137–252, 1980, Conjecture 1.2.10) who predicted its existence. We report on the (ℓ-adic version) of it (L. Lafforgue (Invent Math 147(1):1–241, 2002, Théorème VII.6) in dimension 1, Drinfeld (Moscow Math J 12(3):515–542, 2012, Theorem 1.1) in higher dimension in the smooth case), explain how we used Drinfeld's theorem in the proof of Simpson's integrality conjecture for cohomologically rigid complex local systems in Esnault and Groechenig (Sel Math 24(5):4279–4292, 2018, Theorem 1.1), also why we proceeded like this, and how we combined this idea together with de Jong's conjecture in order to define and obtain an integrability property of the Betti moduli space (de Jong and Esnault, Integrality of the Betti moduli space, 18 pp. Trans. AMS, to appear, Theorem 1.1).

7.1 Motivation on the Complex Side

Given a field automorphism τ of $\mathbb{C}$, we can postcompose the underlying monodromy representation of a complex local system $\mathbb{L}_{\mathbb{C}}$ by τ to define a *conjugate* complex local system $\mathbb{L}_{\mathbb{C}}^{\tau}$. Given a field automorphism σ of $\bar{\mathbb{Q}}_\ell$, which then can only be continuous if this is the identity on $\mathbb{Q}_\ell$, or more generally given a field isomorphism σ between $\bar{\mathbb{Q}}_\ell$ and $\bar{\mathbb{Q}}_{\ell'}$ for some prime number ℓ', the postcomposition of a *continuous* monodromy representation is no longer continuous (unless $\ell = \ell'$ and σ is the identity on $\mathbb{Q}_\ell$), so *we cannot define a conjugate* $\mathbb{L}_\ell^{\sigma}$ of an ℓ-adic local system by this simple postcomposition procedure.

Returning to the complex side, as a consequence of $\pi_1(X(\mathbb{C}), x(\mathbb{C}))$ being finitely generated (we do not even need the finite presentation here), there are finitely many elements

$$(\gamma_1, \ldots, \gamma_s) \in \pi_1(X(\mathbb{C}), x(\mathbb{C}))$$

H. Esnault, *Local Systems in Algebraic-Arithmetic Geometry*, Lecture Notes in Mathematics 2337, https://doi.org/10.1007/978-3-031-40840-3_7

such that the characteristic polynomial map

$$M_B(X,r)^{\square}_{\mathbb{C}} \xrightarrow{\psi^{\square}} N_{\mathbb{C}} = \prod_{i=1}^{s}(\mathbb{A}^{r-1}\times \mathbb{G}_m)_{\mathbb{C}}$$

$$\rho \mapsto (\det(T-\rho(\gamma_1)),\ldots,\det(T-\rho(\gamma_s))),$$

defined in Sect. 6, which factors through

$$M_B(X,r)_{\mathbb{C}} \xrightarrow{\psi} N_{\mathbb{C}},$$

has the property that ψ is a *closed embedding*. The reason is that a semi-simple representation is determined uniquely up to conjugation by the characteristic polynomial function on all $\gamma \in \pi_1(X(\mathbb{C}), x(\mathbb{C}))$. By finite generation of $\pi_1(X(\mathbb{C}), x(\mathbb{C}))$, finitely many suitably chosen ones among them are enough to recognize a semi-simple representation up to conjugation. So denoting $\tau \circ \det(T-\rho(\gamma))$ by $\det(T-\rho(\gamma))^{\tau}$ to unify the notation, we can summarize the discussion as follows:

An automorphism τ of $\mathbb{C}$ yields a commutative diagram

$$(\star)\qquad \begin{array}{ccccc} M_B^{irr}(X,r)(\mathbb{C}) & \xrightarrow{\text{incl.}} & M_B(X,r)(\mathbb{C}) & \xrightarrow{\psi} & N(\mathbb{C}) \\ \downarrow{\scriptstyle \mathbb{L}_{\mathbb{C}}\mapsto\mathbb{L}_{\mathbb{C}}^{\tau}} & & \downarrow{\scriptstyle \mathbb{L}_{\mathbb{C}}\mapsto\mathbb{L}_{\mathbb{C}}^{\tau}} & & \downarrow{\scriptstyle (-)^{\tau}} \\ M_B^{irr}(X,r)(\mathbb{C}) & \xrightarrow{\text{incl.}} & M_B(X,r)(\mathbb{C}) & \xrightarrow[\psi]{} & N(\mathbb{C}) \end{array}$$

The upper script ^{irr} means the irreducible locus.

7.2 Analogy Over a Finite Field

Let us now assume that X is smooth quasi-projective over a finite field $\mathbb{F}_q$. We denote by p the characteristic of $\mathbb{F}_q$. We fix a prime ℓ different from p. We denote by $M_{\ell}^{irr}(X_{\bar{\mathbb{F}}_p}, r)$ the *set* of all rank r ℓ-adic local systems $\mathbb{L}_{\ell}$ defined over $X_{\bar{\mathbb{F}}_p}$ which

- are arithmetic, that is descend to some $X_{\mathbb{F}_{q'}}$ for some finite extension

$$\mathbb{F}_q \subset \mathbb{F}_{q'}(\subset \bar{\mathbb{F}}_p);$$

- • on $X_{\mathbb{F}_{q'}}$ are irreducible over $\bar{\mathbb{Q}}_{\ell}$.

For $r = 1$ the second condition is of course void.

Claim 7.1 $M_\ell^{irr}(X_{\bar{\mathbb{F}}_p}, 1)$ consists of rank 1 ℓ-adic local systems on $X_{\bar{\mathbb{F}}_p}$ of finite order prime to p.

Proof The datum of a rank 1 ℓ-adic local system of finite order prime to p is equivalent to the datum of a Kummer cover of $X_{\bar{\mathbb{F}}_p}$ of order prime to p, which itself is defined over some $\mathbb{F}_{q'}$. This shows one direction. The other one is classical, see [Del80, Théorème 1.3.1], which is proved using Class Field Theory. It is also pleasant to think of it with a swift argument using weights: An element $\mathcal{L} \in M_\ell^{irr}(X_{\bar{\mathbb{F}}_p}, 1)$ lies in $H^1(X_{\bar{\mathbb{F}}_p}, \mathcal{O}_E^\times)$ for some ℓ-adic field E. We write

$$\mathcal{O}_E^\times = k_E^\times \cdot (1 + \mathfrak{m}\mathcal{O}_E)$$

as multiplicative groups via the Teichmüller lift where $\mathfrak{m} \subset \mathcal{O}_E$ is the maximal ideal of $\mathcal{O}_E$. So $\mathcal{L} = \tau \otimes \mathcal{L}'$ where $\tau \in H^1(X_{\bar{\mathbb{F}}_p}, |k_E^\times|) \subset H^1(X_{\bar{\mathbb{F}}_p}, \mathcal{O}_E^\times)$ is a torsion rank 1 local system of order $m \geq 1$ prime to p as $|k_E|^\times$ itself has order prime to p, and $\mathcal{L}' \in H^1(X_{\bar{\mathbb{F}}_p}, (1+\mathfrak{m}\mathcal{O}_E))$. Thus $\mathcal{L}^m = (\mathcal{L}')^m \in H^1(X_{\bar{\mathbb{F}}_p}, (1+\mathfrak{m}\mathcal{O}_E))$ descends to $X_{\mathbb{F}_{q'}}$. The p-adic logarithm $\log_p$ identifies the multiplicative group $(1+\mathfrak{m}\mathcal{O}_E)$ with the additive group $\mathcal{O}_E$. Thus $\log_p(\mathcal{L}^m) \in H^1(X_{\bar{\mathbb{F}}_p}, \mathcal{O}_E) \subset H^1(X_{\bar{\mathbb{F}}_p}, \bar{\mathbb{Q}}_\ell)$ descends to $X_{\mathbb{F}_{q'}}$, which by a weight argument forces $\log_p(\mathcal{L}^m)$ to be equal to 0. This shows the other direction. □

Deligne deduces in [Del80, Proposition 1.3.14] from Claim 7.1 and from the structure of the Tannaka group of an irreducible ℓ-adic local system defined over $X_{\mathbb{F}_{q'}}$ that it is equivalent for the definition of $M_\ell^{irr}(X_{\bar{\mathbb{F}}_p}, r)$ to replace the condition • by the condition

•′ descend to a Weil sheaf to some $X_{\mathbb{F}_{q'}}$ for some finite extension

$$\mathbb{F}_q \subset \mathbb{F}_{q'} (\subset \bar{\mathbb{F}}_p).$$

He shows indeed in *loc. cit.* that an irreducible Weil sheaf on $X_{\mathbb{F}_{q'}}$ becomes étale after a twist by a character of $\mathrm{Gal}(\bar{\mathbb{F}}_p/\bar{\mathbb{F}}_{q'})$.

The set $M_\ell^{irr}(X, r)$ is a replacement for $M_B^{irr}(X, r)(\mathbb{C})$ on the left side of $(\star)$. While over $\mathbb{C}$ we have a scheme structure underlying this set, over $\bar{\mathbb{F}}_p$ we just consider the set. A replacement for $M_B(X, r)(\mathbb{C})$ should take into account the various factors and on each factor, the various twists by a character of $\mathrm{Gal}(\bar{\mathbb{F}}_p/\bar{\mathbb{F}}_{q'})$. This is the reason why in $(\star)$ we restrict the discussion to $M_B^{irr}(X, r)(\mathbb{C})$.

We give ourselves an abstract field isomorphism $\sigma : \bar{\mathbb{Q}}_\ell \xrightarrow{\cong} \bar{\mathbb{Q}}_{\ell'}$. The only "continuous"' information it contains is that it sends a number field $K \subset \bar{\mathbb{Q}}_\ell$ to another one $K^\sigma \subset \bar{\mathbb{Q}}_{\ell'}$. So the right vertical arrow $(-)^\sigma$ makes sense only on a $\gamma \in \pi_1(X_{\mathbb{F}_{q'}}, x_{\bar{\mathbb{F}}_p})$, where $\mathbb{F}_q \subset \mathbb{F}_{q'} (\subset \bar{\mathbb{F}}_p)$ is a finite extension, which has the property that *the characteristic polynomial of γ has values in a number* *field.* Furthermore, we wish the set of such γ to be able to *recognize completely* $M_\ell^{irr}(X,r)$ as ψ does over $\mathbb{C}$. We know the following fact.

Fact 7.2 The set of *conjugacy classes of the Frobenii at all closed points* $|X|$ *of* X has those two properties by [Laf02, Théorème 7.6] and Čebotarev's theorem.

So we set $N^\infty = \prod_{|X|}(\mathbb{A}^{r-1} \times \mathbb{G}_m)$ and ψ^∞ for the characteristic polynomial map on those Frobenii of all closed points. Then *Deligne's companion conjecture may be visualized on the diagram*

$$(\star\star) \quad \begin{array}{ccc} M_\ell^{irr}(X,r) & \xrightarrow{\psi^\infty} & N^\infty(\bar{\mathbb{Q}} \subset \bar{\mathbb{Q}}_\ell) \\ \Big\downarrow{\scriptstyle ?\exists\, \mathbb{L}_\ell \mapsto \mathbb{L}_\ell^\sigma} & & \Big\downarrow{\scriptstyle (-)^\sigma} \\ M_{\ell'}^{irr}(X,r) & \xrightarrow[\psi^\infty]{} & N^\infty(\bar{\mathbb{Q}} \subset \bar{\mathbb{Q}}_{\ell'}) \end{array}$$

The wished $\mathbb{L}_\ell^\sigma$ is called the *companion* of $\mathbb{L}_\ell$ for σ. See the ℓ-adic part in [Del80, Conjecture 1.2.10] for Deligne's precise formulation. There are also p-to-ℓ-companions which we do not discuss, see [AE19], also [Ked22].

7.3 Geometricity

The formulation $(\star\star)$ hides certain properties. For the fact that the eigenvalues of Frobenius on our irreducible ℓ-adic sheaves are *algebraic numbers*, we quote *loc. cit.* which postdates Deligne's conjecture. This is because if X is a smooth curve over $\mathbb{F}_q$, L. Lafforgue proves as a corollary of the Langlands program that an $\mathbb{L}_\ell$ as before is even *pure*. In fact Deligne in [Del80] and Deligne-Beilinson-Bernstein-Gabber in [BBDG82] prove that geometric local systems are pure or mixed, and Lafforgue proves geometricity on curves. Up to now, *we do not know geometricity* of arithmetic local systems over X smooth quasi-projective of higher dimension over $\mathbb{F}_q$. But we know thanks to Drinfeld [Dri12] how to extend the existence of companions from curves to such an X. Let us remark that even on curves, *we cannot* lift the geometricity property from $\bar{\mathbb{F}}_p$ to $\mathbb{C}$, see Sect. 6.6.

Nonetheless, in view of Simpson's geometricity conjecture [Sim92, p. 9] for irreducible rigid complex local systems, the fact that, in absence of geometricity, Drinfeld shows how to prove some integrality property for ℓ-adic local systems in the sense that once we have a geometrically irreducible ℓ-adic local system, we have all geometrically irreducible ℓ'-adic local systems with the "same" characteristic polynomials at closed points, for any $\ell' \neq \ell$, was the philosophical reason behind our proof in [EG18, Theorem 1.1].

7.4 On Drinfeld's Proof

We sketch some aspects of it and refer to the original article [Dri12] for the complete proof. First let us comment that Drinfeld's proof unfortunately does not solve the conjecture on X normal over $\mathbb{F}_q$ as originally formulated by Deligne, but "only" (if this is possible to say "only" here) on X smooth. One problem is that the restriction of simple arithmetic local systems on a non-normal subvariety is not understood, in particular when/whether it is semi-simple. Not even for complex varieties is this problem of preservation of semi-simplicity by pullback understood, see on this [Moc07, Theorem 25.30], where there is a typo, (the proof truly does assume that the domain and the target are smooth and it can be extended to normal varieties). See also [dJE22, Section 7.3] for an arithmetic proof of the semi-simplicity statement [Moc07, Theorem 25.30] for normal varieties, and [D'Ad20, Theorem 1.2.2] for further studies in the non-normal case.

We fix $\mathbb{L}_\ell \in M_\ell^{irr}(X, r)$, an isomorphism $\sigma : \bar{\mathbb{Q}}_\ell \xrightarrow{\cong} \bar{\mathbb{Q}}_{\ell'}$. By Deligne's Theorem [Del12, Théorème 3.1], $\psi^\infty(\mathbb{L}_\ell) \in N^\infty(\mathcal{O})$ where $\mathcal{O}$ is a number ring in $\bar{\mathbb{Q}}_\ell$. This fixes the ℓ-adic completion $\mathcal{O}_E \subset \bar{\mathbb{Q}}_\ell$ where E is an ℓ-adic field, together with the number ring $\sigma(\mathcal{O}) \subset \bar{\mathbb{Q}}_{\ell'}$ and its completion to an ℓ'-adic ring $A \subset \bar{\mathbb{Q}}_{\ell'}$. We denote by $\mathfrak{m}_E \subset \mathcal{O}_E$, $\mathfrak{m} \subset A$ the maximal ideals. The starting point of the proof is the datum (via [Laf02], *loc. cit.*) for any smooth curve $C/\mathbb{F}_q$, any morphism $f_C : C \to X$ which is generically an embedding, of a semi-simple ℓ'-adic local system $\mathbb{L}_{A,C}$ defined over A on C with the property that on intersection points $C \times_X C'$ the pull-backs from C and C' have the same characteristic polynomials. Moreover there are such curves C so that $\mathbb{L}_{A,C}$ is geometrically irreducible. Those local systems $\mathbb{L}_{A,C}$ are simply obtained as the A-forms of the companions $\mathbb{L}_\ell|_C^\sigma$ of $\mathbb{L}_\ell|_C$.

The goal is to construct an irreducible A-adic local system $\mathbb{L}_A$ on X with restrictions on those curves C isomorphic to $\mathbb{L}_{A,C}$. This will be the companion $\mathbb{L}_\ell^\sigma$ of $\mathbb{L}_\ell$ for σ, forgetting in the notation that it is definable over A. If we can do this, by Čebotarev's density theorem, $\mathbb{L}_A$ is unique.

7.4.1 Reductions

To this aim, we first construct a companion on some smooth variety X' endowed with a finite étale Galois cover $h : X' \to X$ with Galois group Γ. In this situation there is a smooth geometrically connected curve $\iota : C \to X$ such that the composed morphism

$$q : \pi_1(C, x) \xrightarrow{\iota_*} \pi_1(X, x) \xrightarrow{q_0} \Gamma$$

is surjective, where $x \to C$ is a geometric point, see e.g. [EK12, Proposition B.1] and references therein. Write $h_C : C' = C \times_X X' \to C$ and $\iota' : C' \to X'$ for the pulled back morphisms.

Claim 7.3 Let $\mathbb{L}_{X'}$ be an ℓ-adic local system, such that $\iota'^*\mathbb{L}_{X'} =: \mathbb{L}_{C'}$ descends to an ℓ-adic local system $\mathbb{L}_C$ on C, i.e. $\mathbb{L}_{C'} = h_C^*\mathbb{L}_C$. Then $\mathbb{L}_{X'}$ descends to an ℓ-adic local system $\mathbb{L}_X$ on X, i.e. $\mathbb{L}_{X'} = h^*\mathbb{L}_X$.

Remark 7.4 Drinfeld does not directly write Claim 7.3. The reduction is hidden in his proof of the boundedness result below. We also remark, even if this is useless here, that if X was defined over $\mathbb{C}$ instead, then ι would exist as well and the Claim would be true with complex local systems replacing ℓ-adic ones, with the same proof.

Proof By conjugating a representation $\rho_{X'} : \pi_1(X', x) \to GL_r(\bar{\mathbb{Q}}_\ell)$ corresponding to $\mathbb{L}_X$ by an element in $GL_r(\bar{\mathbb{Q}}_\ell)$, we may assume that $(\iota')^*\rho_{X'} = (h')^*\rho_C$ where $\rho_C : \pi_1(C, x) \to GL_r(\bar{\mathbb{Q}}_\ell)$ is a representation corresponding to $\mathbb{L}_C$. The surjectivity of q implies the surjectivity of

$$\mathrm{Ker}\big(\pi_1(C', x) \to \pi_1(X', x))\big) \to \mathrm{Ker}\big(\pi_1(C, x) \to \pi_1(X, x))\big)$$

which implies that t ρ_C factors through $\bar{\rho}_C : \bar{\pi}_1(C, x) \to GL_r(\bar{\mathbb{Q}}_\ell)$ where

$$\bar{\pi}_1(C, x) = \iota_*\pi_1(C, x) \subset \pi_1(X, x).$$

For each element $\gamma \in \Gamma$ we choose an element $c_\gamma \in \bar{\pi}_1(C, x)$ such that $q_0(c_\gamma) = \gamma$. Then any element $g \in \pi_1(X, x)$ can be uniquely written as $g = c_\gamma \cdot h_*(\alpha)$ for some $\gamma \in \Gamma$ and $\alpha \in \pi_1(X', x)$. We set

$$\rho_X(g) = \bar{\rho}_C(c_\gamma) \cdot \rho_{X'}(\alpha).$$

It is straightforward to check that ρ_X defines a continuous representation.

□

We construct h as follows. First we "tamify" $\mathbb{L}_X$: We define the finite étale Galois cover $h_0 : Y_0 \to X$ associated to the residual representation in $GL_r(\mathcal{O}_E/\mathfrak{m}_E)$ if $\ell \geq 3$ else to $GL_r(\mathcal{O}_E/\mathfrak{m}_E^2)$ if $\ell = 2$. Then, after replacing $\mathbb{F}_q$ by a finite extension $\mathbb{F}_{q'}$, there is a Zariski dense open on $U \subset Y_0 \otimes_{\mathbb{F}_q} \mathbb{F}_{q'}$ and an elementary fibration on U in the sense of Artin [SGA4, Exposé XI, Proposition 3.3]. This means that there is a fibration $f : U \to S$ of relative dimension 1, which is smooth, with a relative compactification $\iota : U \hookrightarrow \bar{U}$ such that $\bar{U} \setminus U \to S$ is finite étale. We take $h : Y \to X$ to be the Galois hull of the composition of h_0 with $Y_0 \otimes_{\mathbb{F}_q} \mathbb{F}_{q'} \to Y_0$. The ℓ-adic local system $h^*\mathbb{L}_\ell$ is now semi-simple as an étale sheaf, all of its summands are tame. So we may assume $X = Y$, $\mathbb{L}_\ell$ is tame and there is an elementary fibration on a dense open U of X.

We can replace X by $U \subset X$. This follows from the ramification results by Kerz-Schmidt [KS09, Lemma 2.1] and [KS10, Lemma 2.9] based on Wiesend's work [Wie06]: let $\mathbb{L}^\sigma_{\ell,U}$ be the companion of $\mathbb{L}_\ell$ on U. If it ramifies on $X \setminus U$ then there is a curve $C \to X$ such that $\mathbb{L}^\sigma_{\ell,U}|_{C \times_X U}$ is ramified along $C \setminus C \times_X U$. By [Del72, Théorème 9.8], companions on curves preserve ramification. This is a contradiction.

So we may assume that X admits an elementary fibration $f : X \to S$ and $\mathbb{L}_\ell$ is tame. Applying again Claim 7.3 we may assume that the elementary fibration has a section $\sigma : S \to X$, and in addition, we can shrink S to a dense open and X to the inverse image.

7.4.2 Boundedness

The main step [Dri12, Proposition 2.12] in Drinfeld's proof is a boundedness result, which is a generalization to the non-abelian case of Wiesend's result [Wie06, Proposition 17]. After possibly shrinking S to a dense open and X to the inverse image, Drinfeld constructs an infinite sequence

$$\ldots \to X_{n+1} \to X_n \to \ldots \to X_1 \to X$$

of finite étale covers, such that

(i) for any $n \in \mathbb{N}_{>0}$, for any $C \to X$ as above, the $G_n = GL_r(A/\mathfrak{m}^n)$-torsor defined by $\mathbb{L}_{A,C\times_X X_n}$ is trivial;
(ii) for $x \to X$ a geometric point, setting $H_n = \pi_1(X_n, x)$, the group $\cap_n H_n \subset \pi_1(X, x)$ contains a subgroup H which is closed and normal in $\pi_1(X, x)$ such that $\pi_1(X, x))/H$ is almost pro-ℓ, that is an extension of a finite group by a pro-ℓ-group.

Property (ii) does not depend on the choice of x.

If $\dim(X) = 1$, we apply the existence of companions on curves, then H is defined to be the kernel of the monodromy representation, H_n is defined to be the kernel of the truncated representation $\pi_1(X_{\bar{\mathbb{F}}_q}, x_{\bar{\mathbb{F}}_q}) \to GL_r(A) \to G_n$ so $H = \cap_n H_n$, and $X_n \to X$ is the finite Galois étale cover defined by H_n.

We assume $\dim(X) \geq 2$ and that X is an elementary fibration denoted $f : X \to S$, so $X \to S$ is a relative curve with a good relative compactification $\iota : X \hookrightarrow \bar{X}$ above S, such that $\bar{X} \setminus X \to S$ is finite étale, with a section $\sigma : S \to X$.

7.4.3 Moduli

We are grateful to Dustin Clausen for a thorough discussion on Claim 7.5. The idea of the proof presented here, based on Drinfeld's idea, is due to him.

Claim 7.5 ([Dri12], Lemma 3.1, Proof) Replacing S by a dense open, there is a tower of finite étale surjective morphisms

$$\ldots \to T_{n+1} \to T_n \to \ldots \to T_1 \to S$$

such that T_n is the fine moduli space of tame G_n-torsors on X which are trivial above the section $\sigma(S)$.

Proof We consider the three étale sheaves in groupoids over S. For $S' \to S$

(1) $\mathcal{F}_n(S')$ is the groupoid of tame G_n-torsors $X' \to X$ equipped with a trivialization over the section $\sigma' = \sigma \times_S S' : S' \to X' = X \times_S S'$;
(2) $\mathcal{G}_n(S')$ is the groupoid of tame G_n-torsors over X';
(3) $\mathcal{H}_n$ is the groupoid of G_n-torsors over S'.

By [SGA1, Exposé XIII, Corollaire 2.7], $\mathcal{H}_n$ and $\mathcal{G}_n$ are 1-constructible and compatible with base change. On the other hand, $\mathcal{F}_n$ is the equalizer of the restriction map $(\sigma')^* : \mathcal{G}_n(S') \to \mathcal{H}_n(S')$ with the map $\mathcal{G}_n(S') \to \mathcal{H}_n(S')$ assigning to any G_n-torsor on X' the trivial one on S'. Thus by [SGA1, Exposé XIII, Lemme 3.1.1], $\mathcal{F}_n$ is 1-constructible and compatible with base change as well. In addition, $\mathcal{F}_n$, viewed as a presheaf, is already a sheaf as the trivialisation on the section forces $\mathrm{Aut}(\mathcal{F}_n(S')) = \{1\}$. So there are dense open subschemes $S_{n+1} \subset S_n \subset S$ such that $\mathcal{F}_n$ restricted to S_n is represented by a scheme T_n which is finite étale over S_n. By the reductions we may assume $S_1 = S$. So tame G_1-torsors on X' together with a trivialization over σ' are identified with S'-points of $T_1 \to S$.

We explain now why $\mathcal{F}_{n+1} \to \mathcal{F}_n$ is étale finite surjective for $n \geq 1$. We denote by V_{n+1} the kernel of $G_{n+1} \to G_n$. It is a commutative group, in fact a finite dimensional vector space over $\mathbb{F}_{\ell'}$. It is not central, which is to say that the induced homomorphism $\chi_n : G_n \twoheadrightarrow \Gamma_n \subset \mathrm{Aut}(V_{n+1})$ defined by lifting to G_{n+1} and conjugating on V_{n+1} is not trivial. The obstruction to lift any torsor $P_n \in H^1(X, G_n)$ to a G_{n+1}-torsor on X lies in $H^2(X, V_{n+1}^{\chi_n})$, where $V_{\ell+1}^{\chi_n}$ if the $V_{\ell+1}$-torsor defined by pushing P_n along χ_n. By cohomological dimension of f there is an exact sequence

$$H^2(S, f_* V_{n+1}^{\chi_n}) \to H^2(X, V_{n+1}^{\chi_n}) \to H^1(S, R^1 f_* V_{n+1}^{\chi_n}).$$

As ℓ' is prime to p, all V_{n+1} are tame, thus there are only finitely many such local systems $V_{n+1}^{\chi_n}$. As f is an elementary fibration, $V_{n+1}^{\chi_n}$ is tame, the constructible sheaves $R^i f_* V_{n+1}^{\chi_n}$ for $i \geq 0$ are local systems. Thus there is a finite étale cover $T_\circ \to S$, defining $b_\circ : X_{T_\circ} = X \times_S T_\circ \to X$, such that the image of $H^2(X, V_{n+1}^{\chi_n})$ in $H^2(X_{T_\circ}, b_\circ^* V_{n+1}^{\chi_n})$ is zero.

Once a G_n-torsor on X lifts to a G_{n+1}-torsor on $X_{T_\circ}$, to trivialize it on the section $\sigma_{T_\circ} = \sigma \otimes_S T_\circ$, we argue similarly: there are finitely many V_{n+1}-torsors on $T_\circ$ so after a finite étale base change $T \to T_\circ$ defining $b : X_T = X \times_S T \to X$, they are all trivial. We conclude that for any $S' \to S$ and any S'-point $S' \to T_n$ over S, there is a lift $S \times_S T \to T_{n+1}$ thus $T_{n+1} \to T_n$ is finite étale for $n \geq 1$. This finishes the proof.

□

So there is a universal G_n-torsor over $T_n \times_S X$ with Weil restriction $Y_n \to X$ to X. This yields a tower of finite étale surjective morphisms

$$\ldots \to Y_{n+1} \to Y_n \to \ldots \to Y_1 \to X.$$

On the other hand, by the induction assumption, we have a tower

$$\ldots \to S_{n+1} \to S_n \to \ldots \to S_1 \to S$$

which verifies (i) and (ii) for X replaced by S. Then $X_n = Y_n \times_S S_n$ is inserted in a tower

$$\ldots \to X_{n+1} \to X_n \to \ldots \to X_1 \to X$$

of finite étale surjective morphisms. The claim is that this is the sought-after tower.

Property (i) is clear as by definition, for any geometric point $s \to S$,

$$\mathrm{Hom}_S(s, Y_n) = \{\text{tame } G_n\text{-torsors over } U_s\}$$

so for such a geometric point s, $(Y_n)_s \to X_s$ is the universal finite étale cover which trivializes every tame G_n-torsor over X_s.

We address Property (ii). Let $\eta \to S$ be a generic point. We take x to be the image by the section σ of a geometric point $\bar{\eta}$ above η. As X is normal, $\pi_1^t(X_\eta, x) \to \pi_1^t(X, x)$ is surjective, so we can replace X with X_η for (ii). This yields the homotopy exact sequence of fundamental groups

$$1 \to K := \pi_1^t(X_{\bar{\eta}}, x) \to \pi_1^t(X_\eta, x) \to \Gamma := \pi_1(\eta, x) \to 0$$

where the upper index ^{t} indicates tameness with respect to ι and the lower index indicates the fibre. We define

$$H_n = \mathrm{Ker}\big(\pi_1^t(X_\eta, x) \to \mathrm{Aut}((X_n)_x)\big).$$

Then H_n is inserted in an exact sequence

$$1 \to K_n \to H_n \to \Gamma_n \to 1$$

where

$$K_n = H_n \cap K = \cap_\rho \mathrm{Ker}\big(\pi_1^t(X_{\bar{\eta}}, x) \xrightarrow{\rho} G_n\big)$$

$$\Gamma_n = \mathrm{Ker}\big(\Gamma \to \mathrm{Aut} K/K_n\big).$$

Since K is topologically finitely generated, there are finitely many ρ's in the definition of K_n thus $K/\cap_n K_n$ is almost pro-ℓ. Finally, $\Gamma/\cap_n \Gamma_n$ is a quotient of $\mathrm{Im}\big(\Gamma \to \mathrm{Aut}(K/\cap_n K_n)\big)$ while $\mathrm{Aut}(K/\cap_n K_n)$ is almost pro-ℓ as well. Indeed an automorphism has to respect the maximal pro-ℓ sub and in it the kernel to its first $\mathbb{F}_\ell$-homology, which itself is pro-ℓ and of finite index. This proves (ii) for $H = \cap_n H_n$ and finishes the proof of this first part for X with an elementary fibration $X \to S$.

7.4.4 Gluing

The main point is then to prove that the preceding property suffices to glue the $\mathbb{L}^\sigma_{\ell,C}$. The proof follows an idea of Kerz, see [Dri12, Section 4], Drinfeld himself initially had a more difficult argument. First we do a Lefschetz type argument to find a curve $\varphi : C \to X$ containing x such that $\varphi_* : \pi_1(C, x) \to \pi_1(X, x)/H$ is surjective. This is possible as the surjectivity is recognized on the finite quotient which is the extension of the residual quotient of $\pi_1(X)/H$ (if $\ell \geq 3$, if $\ell = 2$ we go mod $\mathfrak{m}^2$ here) by the maximal $\mathbb{F}_\ell$-vector space quotient, see [EK11, Lemma 8.2] (see also above where a similar property was already used in order to show that $\mathrm{Aut}(K/\cap_n K_n)$ is almost pro-ℓ). By (i) and (ii) the restriction of $\mathbb{L}^\sigma_{\ell,C}$ to $\mathrm{Ker}(\varphi_*)$ is unipotent, but it has to be semi-simple as well as $\mathrm{Ker}(\varphi_*) \subset \pi_1(C, x)$ is normal and $\mathbb{L}^\sigma_{\ell,C}$ itself is semi-simple. Thus the monodromy group of $\mathbb{L}^\sigma_{\ell,C}$ is a quotient of $\pi_1(C, x)/\mathrm{Ker}(\varphi_*) = \pi_1(X, x)/H$, and this *defines* a representation $\pi_1(X, x) \to GL_r(A)$ *which a priori depends on* C. Denote by $\mathbb{L}_{\ell',X}$ the underlying local system. By definition its restriction to C is equal to $\mathbb{L}^\sigma_{\ell,C}$. To compute the restriction of $\mathbb{L}_{\ell',X}$ to all closed points, we take other curves $\varphi_{C'} : C' \to X$ with the same property so they fill in all the closed points. If we make sure that C and C' meet in sufficiently many points, then $(\mathbb{L}_{\ell',X}|_C = \mathbb{L}^\sigma_{\ell,C}, \mathbb{L}_{\ell',X}|_{C'}, \mathbb{L}^\sigma_{\ell,C'})$ are recognized by their value on the intersection points by [Fal83, Satz 5]. This finishes the proof for X with an elementary fibration $X \to S$.

7.5 Cohomologically Rigid Local Systems Are Integral, See [EG18], Theorem 1.1

We sketch the proof with Michael Groechenig in *loc.cit.*: let X be a smooth quasi-projective variety defined over the field of complex numbers, let $T_i \subset GL_r(\mathbb{C})$ be conjugacy classes of quasi-unipotent matrices, and $\mathcal{L}$ be a torsion rank 1 local system. We denote by $M_B(X, r, \mathcal{L}, T_i^{ss})$ the Betti moduli of rank r local systems with determinant $\mathcal{L}$ and semi-simplification of the monodromy at infinity being the semi-simplification T_i^{ss} of T_i. We refer to Sect. 7.3 for the heuristic of the following theorem.

Theorem 7.6 *Irreducible cohomologically rigid local systems in $M_B(X, r, \mathcal{L}, T_i^{ss})$ are integral.*

Proof Let X be a smooth quasi-projective variety over $\mathbb{C}$, fix a torsion character $\mathcal{L}$ and quasi-unipotent conjugacy classes $T_i \subset GL_r(\mathbb{C})$. Let $\mathbb{L}_\mathbb{C}$ be a point of $M_B(X, r, \mathcal{L}, T_i)(\mathbb{C})$ which is isolated. It is defined by a representation $\rho_\mathbb{C}$: $\pi_1(X(\mathbb{C}), x(\mathbb{C})) \to GL_r(A)$ where A is a ring of finite type over $\mathbb{Z}$. Let $a : A \to \bar{\mathbb{Z}}_\ell$ be a $\bar{\mathbb{Z}}_\ell$-point of A. It defines an ℓ-adic local system $\mathbb{L}_{\ell,\mathbb{C}}$ by the factorization

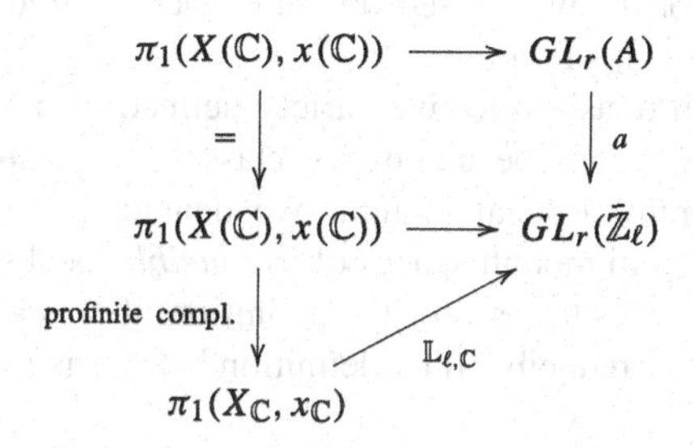

Using the notation of 6.2, for a closed point $s \in |S|$ of characteristic prime to the order of the residual representation of $\mathbb{L}_{\ell,\mathbb{C}}$, the order of $\mathcal{L}$ and of the eigenvalues of the monodromies at infinity, and integral for the (finitely many) cohomologically rigid local systems, the following properties hold.

(1) $\mathbb{L}_{\ell,\mathbb{C}}$ descends to $\mathbb{L}_{\ell,s'}$ [Sim92, Theorem 4] for some $s' \to s$ finite below $\bar{s}$ (one has to complete the argument to take care of the conditions at infinity) keeping $\mathcal{L}$ and T_i^{ss} [Del73, Section 1.1.10];
(2) The companions $\mathbb{L}^\sigma_{\ell,s'}$ still have determinant $\mathcal{L}$ and the semi-simplification of the monodromies at infinity is the one of the T_i^{ss}'s [Del72, Théorème 9.8];
(3) The intersection cohomology verifies:

$$IH^1(X, \mathcal{E}nd^0(\mathbb{L}_{\ell,\bar{s}})) = 0 = IH^1(X, \mathcal{E}nd^0(\mathbb{L}^\sigma_{\ell,\bar{s}}))$$

by the weight argument [EG18, Proof of Theorem 1.1];
(4) So $\mathbb{L}_{\ell,\mathbb{C}} \mapsto sp^{-1}_{\mathbb{C},\bar{s}}(\mathbb{L}^\sigma_{\ell,\bar{s}}))$ is a bijection from the set of cohomologically rigid ℓ-adic local systems to the set of cohomologically rigid ℓ'-adic local systems.

So there cannot be a non-integral place ℓ'. This finishes the proof. □

Remarks 7.7

(1) The "same" proof works if GL_r is replaced by a reductive group $G \subset GL_r$, see [KP22] by Klevdal-Patrikis.
(2) It is not the case that all rigid local systems are cohomologically rigid, see [dJEG22]. In the present state of knowledge, the examples known are all cohomologically rigid in a smaller reductive group $G \subset GL_r$ as in (1). However, I have no philosophical argument why rigid local systems should be cohomologically rigid in some smaller reductive group $G \subset GL_r$ or why not.

7.6 Integrality of the Whole Betti Moduli Space, See [dJE22], Theorem 1.1

In Theorem 6.15 we proved using de Jong's conjecture that weakly arithmetic local systems are dense in their Betti moduli. In Theorem 7.6 we proved using the existence of the ℓ-adic companions that cohomologically rigid local systems are integral. Combining the two essential ideas in those theorems we define a (weak) notion of integrality for the whole Betti moduli space of irreducible local systems and prove the following theorem.

Let X be a smooth quasi-projective variety defined over the field of complex numbers, let $T_i \subset GL_r(\mathbb{C})$ be conjugacy classes of quasi-unipotent matrices, and $\mathcal{L}$ be a torsion rank 1 local system. We denote by $M_B^{irr}(X, r, \mathcal{L}, T_i^{ss}) \subset M_B(X, r, \mathcal{L}, T_i^{ss})$ the Betti moduli spaces of *irreducible* local systems. See [dJE22, Section 2.1] and references therein for the definition: if $\mathbb{L}$ is a point and is defined over a ring A, then $\mathbb{L}$ is irreducible if by definition $\mathbb{L} \otimes_A \kappa$ is irreducible for all field valued points $A \to \kappa$.

Theorem 7.8 *If there is one complex point of $M_B(X, r, \mathcal{L}, T_i)$ which is irreducible, then for all prime numbers ℓ, there is an irreducible ℓ-adic local system of rank r with determinant $\mathcal{L}$ and monodromies at infinity being conjugate to T_i up to semi-simplification.*

Proof The existence of a complex point $\mathbb{L}_\mathbb{C}$ which is irreducible is equivalent to saying that the morphism $M^{irr}(X, r, \mathcal{L}, T_i^{ss}) \to \mathrm{Spec}(\mathcal{O}_K)$ is dominant, where K is the number field over which $\mathcal{L}$ and the T_i^{ss} are defined. By generic smoothness, we find a closed point z of characteristic $\ell > 0$ on $M_B^{irr}(X, r, \mathcal{L}, T_i^{ss})$ which is smooth for the morphism $M^{irr}(X, r, \mathcal{L}, T_i^{ss})_{red} \to \mathrm{Spec}(\mathcal{O}_K)$. We choose $s \in S$ a closed point of the scheme of definition of a good compactification $X \hookrightarrow \bar{X}$, $\mathcal{L}$ and T_i such that z descends to $X_{\bar{s}}$. Applying now the argument as in the proof of Theorem 6.15, there is on $X_{\bar{s}}$ an irreducible rank r arithmetic $\bar{\mathbb{Q}}_\ell$-local system with z as a residual local system, and with monodromies at infinity which are in the conjugacy class of T_i up to semi-simplification. Choosing $\sigma : \bar{\mathbb{Q}}_\ell \cong \bar{\mathbb{Q}}_{\ell'}$ now as in the proof of Theorem 7.6, the σ-companion on the same $X_{\bar{s}}$ has the right determinant and monodromies at infinity. We pull it back to $X_\mathbb{C}$ via $sp_{\mathbb{C},\bar{s}}$ as in the proof of Theorem 7.6. This finishes the proof.

□

Remark 7.9 We can enhance Theorem 7.8 in various ways. Rather than assuming there is *one* irreducible complex local system, we can ask for *infinitely many* pairwise non-isomorphic $\mathbb{L}_\mathbb{C}$, then the conclusion follows, there are infinitely many such irreducible $\mathbb{L}_\ell$ for all ℓ [dJE22, Theorem 1.4]. We can also ask for the algebraic monodromy of $\mathbb{L}_\mathbb{C}$ to e.g. contain $SL_r(\mathbb{C})$, then the conclusion follows for $\mathbb{L}_\ell$ [dJE22, Theorem 1.5] etc. But the main point which would be wishful to understand is not understood: with this method initiated in [EG18] to use companions to obtain integrality, we do not know whether on $X_{\bar{s}}$ in characteristic p where we perform this construction, the "dimension of the components" of $M_B(X, r, \mathcal{L}, T_i^{ss})$ over $\bar{\mathbb{Z}}_\ell$ is preserved as we go over to $\bar{\mathbb{Z}}_{\ell'}$ via the companions. Stated like this it makes no proper sense. One should specify that at ℓ we complete $M_B^{irr}(X, r, \mathcal{L}, T_i^{ss})$ at a closed point which is smooth for $M_B^{irr}(X, r, \mathcal{L}, T_i^{ss}) \to \mathrm{Spec}(\mathcal{O}_K)$, then the component containing z is well defined and had a dimension d. One could dream that the σ-companion is also on a dimension d component of $M_B(X, r, \mathcal{L}, T_i^{ss})$. For example for $d = 0$ this would prove entirely Simpson's integrality conjecture.

7.7 Obstruction

Forgetting the conditions at infinity, if we fix a finitely presented group Γ, a natural number $r \geq 1$, and a rank 1 torsion complex character $\chi : \Gamma \to \mathbb{C}^\times$, we posed in [dJE22, Definition 1.2] the following definition.

Definition 7.10 Γ has the weak integrality property with respect to (r, χ) if, assuming there is an irreducible representation $\rho : \Gamma \to GL_r(\mathbb{C})$ with determinant χ, then for any prime number ℓ, there is a representation $\rho_\ell : \Gamma \to GL_r(\bar{\mathbb{Z}}_\ell)$ with determinant χ which is irreducible over $\bar{\mathbb{Q}}_\ell$.

Then, by the work of Becker-Breuillard-Varjù [BBV23], there are pairs (Γ, χ) which do *not have the weak integrality property*. One example is given by Γ being the group spanned by two letters a, b with one relation $b^2a^2 = a^2b$, χ is the trivial character and $r = 2$. The example is not integral at $\ell = 2$, see [dJE22, Example 7.4]. Given Theorem 7.8, one easily computes that if Γ is the topological fundamental group of $X(\mathbb{C})$ for X smooth quasi-projective over $\mathbb{C}$ and χ is any torsion character, then it has the weak integrality property (see [dJE22, Theorem 1.3])). So we obtain the following theorem.

Theorem 7.11 *Let Γ be a finitely presented group, χ be a torsion character of Γ, $r \geq 1$ a natural number. If Γ does not have the integrality property with respect to (r, χ), Γ cannot be the topological fundamental group of a smooth complex quasi-projective variety.*

The proof relies on Theorem 7.8. This so defined obstruction is of course by no means coming from easy theorems, at least in the present state of knowledge. As explained before, Theorem 7.8 relies both on the Langlands duality for a curve over a finite field and on the geometric Langlands duality. But it has the advantage that in the definition of the obstruction we do not need to specify which elements of Γ are supposed to come from the local fundamental groups at infinity: none of those can have this property.

Chapter 8
Lecture 8: Rigid Local Systems and F-Isocrystals

Abstract Rigid connections over a variety $X_{\mathbb{C}}$ smooth projective over $\mathbb{C}$, while restricted to the formal p-completion $\hat{X}_W$ of a non-ramified projective p-adic model X_W of $X_{\mathbb{C}}$, yield F-isocrystals. This is proved in Esnault and Groechenig (Acta Math 225(1):103–158, 2020, Theorem 1.6), using the theory of Higgs-de Rham flows on the mod p reduction of X_W. We give here a p-adic proof of this theorem, obtained with Johan de Jong, which relies on the fact that for $p \geq 3$, the Frobenius pull-back of a connection on $\hat{X}_W$ is well defined, whether the p-curvature of the mod p reduction is nilpotent or not. However this proof so far does not give the crystallinity property proved in Esnault and Groechenig (Acta Math 225(1):103–158, 2020, Theorem 5.4) which can be expressed by saying that the p-adic local systems obtained on the p-adic variety $X_{\overline{\mathrm{Frac}(W)}}$ for p large descend to a crystalline local system over $X_{\mathrm{Frac}(W)}$. The version under the strong cohomological rigidity of the same theorem is worked out in Esnault and Groechenig (Frobenius structures and unipotent monodromy at infinity, Preprint 2021, 8 pp.). We defer this discussion to Chap. 9.

8.1 Crystalline Site, Crystals and Isocrystals

We refer to [EG20, Section 2.6] for the details of the definitions, and for the appropriate references.

Given X smooth over a perfect characteristic $p > 0$ field k, we denote by $W = W(k)$ the ring of Witt vectors and by $W_n = W_n(k) = W/p^n$ its n-th truncation. We define the site (X/W_n) with objects the PD-thickenings (U, T, δ) over W_n, where $U \hookrightarrow T$ is a closed embedding defined by an ideal $I \subset \mathcal{O}_T$ on which δ defines a PD-structure over (W_n, pW_n). So for all x_i, $i = 1, \ldots, s$ over (W_n, pW_n) it holds

$$mx_1^{[n_1]} \cdots x_s^{[n_s]} = 0$$

for some powers $n_i, m \in \mathbb{N}$ (see [Ber74, I 1.3.1 ii), p.56]).

H. Esnault, *Local Systems in Algebraic-Arithmetic Geometry*, Lecture Notes in Mathematics 2337, https://doi.org/10.1007/978-3-031-40840-3_8

This yields functors $(X/W_n) \hookrightarrow (X/W_{n+1})$ and the *crystalline site* (X/W) as the colimit over n of the (X/W_n). The Homs are the obvious ones respecting the whole structure. A *crystal* $\mathcal{F}/W$ in coherent modules on X/W is the datum for all (U, T, δ) of a coherent sheaf $\mathcal{F}_T$ so the transition functions are isomorphisms. The category of *isocrystals* has the same objects and the Hom sets, which are abelian groups, are tensored over $\mathbb{Z}$ by $\mathbb{Q}$. For us the main concrete description is [EG20, Theorem 2.19, Corollary 2.20] according to which, *if* $\hat{X}_W$ *is an essentially smooth formal scheme over* W *lifting* X, a crystal is

(i) a flat formal connection $(\hat{E}_W, \hat{\nabla}_W)$ on $\hat{X}_W$;
(ii) such that $(\hat{E}_W, \hat{\nabla}_W) \otimes_W k$ is filtered by subconnections so the associated graded is spanned over a dense open of X by a full set of algebraic solutions (that is its p-curvature is nilpotent, see Sect. 2).

By the classical theorem, (i) implies that that $\hat{E}_K := \hat{E}_W \otimes_W K$ where $K = \mathrm{Frac}(W)$ is locally free. This is not the case for $\hat{E}_W$ even with the condition (ii), see [ES15, 1.3]. It is also not known whether or not one can always find a locally free lattice in $\hat{E}_K$ which is stabilized by $\hat{\nabla}_K$.

8.2 Nilpotent Crystalline Site, Crystals and Isocrystals

We refer to [ES18, Section 1-2] and references in there.

The *nilpotent crystalline site* has objects (U, T, δ) as for the crystalline site, but in addition I itself is locally nilpotent. So for all $x_i,\ i = 1, \dots, s$ over (W_n, pW_n), one has

$$x_1^{[n_1]} \cdots x_s^{[n_s]} = 0$$

for some powers $n_i \in \mathbb{N}$. This yields a continuous functor

$$(X/W)_{Ncrys} \to (X/W)_{crys}$$

of topoi. Crystals and isocrystals are defined similarly but given $\hat{X}_W$ as above, then a crystal on the nilpotent crystalline site is

(i) a flat formal connection $(\hat{E}_W, \hat{\nabla}_W)$ on $\hat{X}_W$.

(See [Ber74, I 3.1. 1 ii) p.56, III 1.3.1 p. 187, IV 1.6.6 p. 248].) There is *no* condition on the mod p reduction.

8.3 The Frobenius Action on the Set of Crystals on the Nilpotent Crystalline Site

As $\hat{X}_W$ is formally smooth over W, the Frobenius of X locally in the Zariski topology lifts to $\hat{X}_W$. The following proposition is due to Berthelot [Ber00, Proposition 2.2.5 a)]. We learned it from de Jong. We are grateful to Kisin for giving us the reference. The proof we present here is taken from [SP, tag/07JH] in which unfortunately the condition ii) is assumed to be fulfilled, but .... not used.

Proposition 8.1 *Let $(\hat{E}_W, \hat{\nabla}_W)$ be a formal flat connection on $\hat{X}_W$. Assume $p \geq 3$. There is a unique formal flat connection denoted by $F^*(\hat{E}_W, \hat{\nabla}_W)$ on $\hat{X}_W$ with the property that if on an open $\hat{U}_W \subset \hat{X}_W$, the Frobenius F of U lifts to F_W, then*

$$F^*(\hat{E}_W, \hat{\nabla}_W)|_{\hat{U}_W} = F_W^*((\hat{E}_W, \hat{\nabla}_W)|_{\hat{U}_W}).$$

***Proof** (loc. cit.)* The goal is to show that if F_W, G_W are two lifts to $\hat{U}_W$ of F on U, there is a commutative diagram

$$\begin{array}{ccc} F_W^*\hat{E}_W & \xrightarrow{\psi} & G_W^*\hat{E}_W \\ \hat{\nabla}\downarrow & & \downarrow\hat{\nabla} \\ \Omega^1_{\hat{U}_W}\hat{\otimes}_{\mathcal{O}_{\hat{U}_W}} F_W^*\hat{E}_W & \xrightarrow{1\otimes\psi} & \Omega^1_{\hat{U}_W}\hat{\otimes}_{\mathcal{O}_{\hat{U}_W}} G_W^*\hat{E}_W \end{array}$$

which is canonical. So then it fulfils the cocycle condition. Here $\hat{\otimes}$ denotes the completed tensor product. So we may assume that $\hat{U}_W$ is étale, finite onto its image, an open of $\hat{\mathbb{A}}^d_W$, where $d = \dim X$, so as to have coordinates $(x_1, \ldots, x_d)$. This defines the derivation $\partial_i \in T_{\hat{U}_W}$ dual to dx_i, acting on $\hat{E}_W|_{\hat{U}_W}$, and the action of the differential operator $\partial_{\underline{k}} = \partial_1^{k_1} \cdots \partial_d^{k_d}$ on $\hat{E}_W|_{\hat{U}_W}$ for all multi-indices $\underline{k} = (k_1, \ldots, k_d)$. We write $F_W^*\hat{E}_W = F_W^{-1}\hat{E}_W \otimes_{F^{-1}\mathcal{O}_{\hat{U}_W}} \mathcal{O}_{\hat{U}_W}$. Then we define

$$\psi(e \otimes 1) = \sum_{\underline{k}} \partial_{\underline{k}}(e) \otimes_{G_W^{-1}\mathcal{O}_{\hat{U}_W}} \prod_{i=1}^{d} (F_W(x_i) - G_W(x_i))^{k_i}/(k_i)!.$$

By definition $(F_W(x_i) - G_W(x_i)) \in p\mathcal{O}(\hat{U}_W)$ so its p-adic valuation is ≥ 1. On the other hand, the p-adic exponential function converges on elements of p-adic valuation $> 1/(p-1)$. So ψ, thus F^*, is defined for $p \geq 3$. This finishes the proof. □

Remark 8.2 If the connection was defined on $X_{\mathcal{O}}$ where $\mathcal{O} \supset W$ is the ring of integers of a p-adic field of degree ≥ 2 over $\mathrm{Frac}(W)$, the same formula with the obvious change of notation would yield $(F_{\mathcal{O}}(x_i) - G_{\mathcal{O}}(x_i)) \in \pi\mathcal{O}(U_W)$ where π is a uniformizer of $\mathcal{O}$. Then the convergence would be violated and we could not write the diagram above. In this case we need condition ii) to define the Frobenius pull-back of $(E_{\mathcal{O}}, \nabla_{\mathcal{O}})$ in this way when the index of ramification of $\mathcal{O}$ lies in $[2, p-1($.

8.4 The Frobenius Induces an Isomorphism on Cohomology in Characteristic 0

We set $K = \mathrm{Frac}(W)$ and keep the notation from the previous paragraph. In addition we set

$$(\hat{E}_W, \hat{\nabla}_W) \otimes_W K = (\hat{E}_K, \hat{\nabla}_K)$$

and abuse notation

$$F^*(\hat{E}_K, \hat{\nabla}_K) := F^*(\hat{E}_W, \hat{\nabla}_W) \otimes_W K.$$

Proposition 8.3 *Let $(\hat{E}_W, \hat{\nabla}_W)$ be a flat formal connection on $\hat{X}_W$ where $p \geq 3$. Then for all cohomological degrees i,*

$$F^* : H^i(\hat{X}_W, (\hat{E}_W, \hat{\nabla}_W)) \otimes_W K \to H^i(\hat{X}_W, F^*(\hat{E}_W, \hat{\nabla}_W)) \otimes_W K$$

is an isomorphism and is compatible with cup-products.

We quote [Ber00, Théorème 4.3.5]. We give below an elementary proof in the style of the one of Proposition 8.1.

Proof By the Mayer-Vietoris spectral sequence it is enough to prove the first part of the proposition on an open $\hat{U}_W$ lifting U on which the Frobenius lifts to $\hat{F}_W$ so as to have an étale map $h : \hat{U}_W \to \hat{\mathbb{A}}^d_W$ finite onto its image, as in Sect. 8.3. To be more precise, we first construct a finite étale morphism $U_k \xrightarrow{h_k} V_k$ onto an affine open $V_k \hookrightarrow \mathbb{A}^d_k$, then a lift of the diagram $U_k \to V_k \hookrightarrow \mathbb{A}^d_k$ to the diagram of formal affine schemes $\hat{U}_W \xrightarrow{h} \hat{V}_W \hookrightarrow \hat{\mathbb{A}}^d_W$.

On $\hat{\mathbb{A}}^d_W$ we choose the lift F_W which is defined by $F_W^*(x_i) = x_i^p$. This defines $\hat{V}'_W := F_W^{-1}\hat{V}_W$. We consider the cartesian diagram

$$\begin{array}{ccc} \hat{U}_W \times_{\hat{V}_W} \hat{V}'_W & \xrightarrow{F'_W} & \hat{U}_W \\ \downarrow{\scriptstyle h\times 1} & & \downarrow{\scriptstyle h} \\ \hat{V}'_W & \xrightarrow{F_W} & \hat{V}_W \end{array}$$

defining F'_W. Both $\hat{U}_W \times_{\hat{V}_W} \hat{V}'_W$ and $\hat{U}_W$ are formal affine schemes with reduction modulo p isomorphic to U_k, so there is an isomorphism of formal schemes

$$\tau : \hat{U}_W \xrightarrow{\cong} \hat{U}_W \times_{\hat{V}_W} \hat{V}'_W$$

over W, see e.g. [Ara01, Théorème 2.1.3], defining

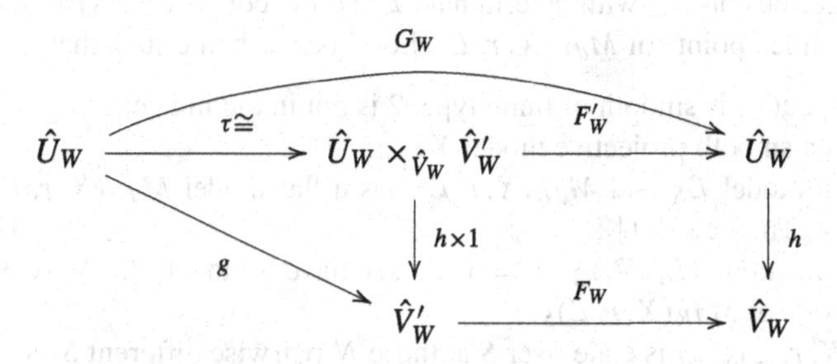

So G_W is a lift of Frobenius. By base change, it holds

$$H^0(\hat{U}_W, G_K^*(\hat{E}_W, \hat{\nabla}_W)) = H^0(\hat{V}'_W, g_* G_K^*(\hat{E}_W, \hat{\nabla}_W)) = H^0(\hat{V}'_W, F_W^* h_*(\hat{E}_W, \hat{\nabla}_W))$$

so we may assume

$$(G_W : \hat{U}_W \to \hat{U}_W) = (F_W : \hat{V}'_W \to \hat{V}_W).$$

Then

$$H^0(\hat{V}'_W, F_W^*(\hat{E}_W, \hat{\nabla}_W)) \otimes_W K = \oplus_{\chi_i, 0 \le n_{ij_i} \le p-1} H^0(\hat{V}_K, (\hat{E}_K, \hat{\nabla}_K) \otimes_{i=1}^d \chi^{n_{ij_i}})$$

where the χ_i are the characters defining the Kummer cover $x_i \mapsto x_i^p$. The only tensor product $\otimes_{i=1}^d \chi^{n_{ij_i}}$ which extends as a formal flat connection on $\hat{V}_W$ is the trivial one, that is the one for which all the $n_{ij_i} = 0,\ i = 1, \ldots, d$. Thus

$$H^i(\hat{V}'_W, F_W^*(\hat{E}_W, \hat{\nabla}_W)) \otimes_W K = H^i(\hat{V}_W, (\hat{E}_W, \hat{\nabla}_W)) \otimes_W K.$$

The second part of the proposition is also checked locally on $\hat{U}_W$ with Frobenius lift F_W, where it is trivial. This finishes the proof. □

8.5 Proof of [EG20], Theorem 1.6

Let $X_{\mathbb{C}}$ be a smooth projective variety. Let $(E_i, \nabla_i)_{\mathbb{C}},\ i = 1, \ldots, N$ be the finitely many irreducible rank r rigid local systems with a fixed torsion determinant $\mathcal{L}$. We denote by the same letter $\mathcal{L}$ the associated rank one connection $(\mathcal{L} \otimes_{\mathbb{C}} \mathcal{O}_{X_{\rm an}}, 1 \otimes d)$. Let $M_{dR}(X, r, \mathcal{L})$ be the de Rham moduli space of irreducible rank r flat connections on $X_{\mathbb{C}}$ with determinant $\mathcal{L}$. So the connections $(E_i, \nabla_i)_{\mathbb{C}}$ are the isolated complex points of $M_{dR}(X, r, \mathcal{L})$. Let S be a scheme such that

(0) $S \to \mathrm{Spec}(\mathbb{Z})$ is smooth of finite type, 2 is not in the image;
(1) $X_{\mathbb{C}}$ has a smooth projective model X_S;
(2) $\mathcal{L}$ has a model $\mathcal{L}_S$ and $M_{dR}(X, r, \mathcal{L})$ has a flat model $M_{dR}(X, r, \mathcal{L})_S$ over S ([Lan14, Theorem 1.1]);
(3) the connections $(E_i, \nabla_i)_{\mathbb{C}},\ i = 1, \ldots, N$ have a model $(E_i, \nabla_i)_S$ defining N S-sections of $M_{dR}(X, r, \mathcal{L})_S$;
(4) $M_{dR}(X, r, \mathcal{L})_{S,red}$ is étale over S at those N pairwise different S-points;
(5) the formal completion of $M_{dR}(X, r, \mathcal{L})_S$ at those sections is flat over S;
(6) the $S \otimes_{\mathbb{Z}} \mathbb{Q}$-modules

$$H^j(X_S, \mathcal{E}nd(E_i, \nabla_i)_S),\ j = 1, 2, i = 1, \ldots, N$$

and

$$K((E_i, \nabla_i))_S = \mathrm{Ker}\big(H^1(X_S, \mathcal{E}nd(E_i, \nabla_i)_S) \xrightarrow{x \mapsto x \cup x} H^2(X_S, \mathcal{E}nd(E_i, \nabla_i)_S)\big)$$

satisfy base change.

The following theorem is [EG20, Theorem 1.6], with a more precise description than in *loc. cit.* of the scheme S on which it holds.

Theorem 8.4 *For any point* $\mathrm{Spec} W \to S$ *where* $W = W(k)$ *and* k *is a perfect characteristic* $p > 0$ *field, for any* $i = 1, \ldots, N$*, the* p*-completion* $(\hat{E}_i, \hat{\nabla}_i)_W$ *of* $(E_i, \nabla_i)_S|_{X_W}$ *is an isocrystal with a Frobenius structure. In particular it has nilpotent* p*-curvature.*

Proof By [GM88, Proposition 4.4] the completion of $K((E, \nabla)_{\mathbb{C}})$ at 0 is isomorphic to the completion at the complex point $(E, \nabla)_{\mathbb{C}}$ of $M_{dR}(X, r, \mathcal{L})$. Grothendieck's comparison isomorphism

$$H^i(X_W, (E_i, \nabla_i)_S|_{X_W}) \xrightarrow{\cong} H^i(\hat{X}_W, (\hat{E}_i, \hat{\nabla}_i)_W)$$

which implies the comparison isomorphism

$$H^i(X_K, (E_i, \nabla_i)_S) = H^i(X_W, (E_i, \nabla_i)_S|_{X_W})_{\mathbb{Q}} \xrightarrow{\cong} H^i(\hat{X}_W, (\hat{E}_i, \hat{\nabla}_i)_W)_{\mathbb{Q}} = H^i(\hat{X}_K, (\hat{E}_i, \hat{\nabla}_i)_K)$$

is compatible with the cup-product. Here $K = \mathrm{Frac}(W)$ and $\hat{X}_K = \hat{X}_W \otimes_W K$. Proposition 8.3 then implies that F^* maps isomorphically the completion of $M_{dR}(X, r, \mathcal{L})_S$ at $(\hat{E}_i, \hat{\nabla}_i)_K$ to the one at $F^*(\hat{E}_i, \hat{\nabla}_i)_K$. Thus the latter is 0-dimensional over K. We conclude

$$F^*(\hat{E}_i, \hat{\nabla}_i)_K = (\hat{E}_{\varphi(i)}, \hat{\nabla}_{\varphi(i)})_K \text{ for some } \varphi(i) \in \{1, \ldots, N\}.$$

Proposition 8.3 for H^0 implies then that $\varphi : \{1, \ldots, N\} \to \{1, \ldots, N\}$ is an injective map, thus is a bijection. Thus there is a natural number M, which divides $N!$, such that

$$\underbrace{\varphi \circ \ldots \circ \varphi}_{\text{M-times}} = \text{Identity}.$$

In other words

$$(F^M)^*(\hat{E}_i, \hat{\nabla}_i)_K = (\hat{E}_i, \hat{\nabla}_i)_K \ \forall i = 1, \ldots, N.$$

This finishes the first part of the proof. As for the second part, we apply [ES18, Proof of Proposition 3.1] which shows that the nilpotency of the p-curvature does not depend on the lattice chosen.

□

Remark 8.5 As compared to [EG20, Theorem 1.6], this proof has the advantage that it is p-adic as opposed to a characteristic $p > 0$ proof. This is more natural as the result is p-adic. It also gives a larger S on which the result holds. It has the disadvantage to juggle with a cohomology which is nowhere studied in details (Proposition 8.3) and, what is the most important point, it does not yield on the nose a Fontaine-Lafaille module on X_W, and its counterpart which is a p-adic crystalline local system on X_K. This is the topic of Chap. 9. The counting argument at the end is of course taken from the proof of [EG20, Theorem 1.6].

Chapter 9
Lecture 9: Rigid Local Systems, Fontaine-Laffaille Modules and Crystalline Local Systems

Abstract As seen in Chap. 8, originally proved with Michael Groechenig in Esnault and Groechenig (Acta Math 225(1):103–158, 2020, Theorem 1.6), rigid connections on $X_{\mathbb{C}}$ smooth projective over $\mathbb{C}$, while restricted to the formal p-completion $\hat{X}_W$ a non-ramified projective p-adic model X_W of $X_{\mathbb{C}}$, yield F-isocrystals. More is true. By showing in Esnault and Groechenig (Acta Math 225(1):103–158, 2020, Theorem 1.6) the existence of a *periodic* Higgs-de Rham flow on the formal connection $(\hat{E}_W, \hat{\nabla}_W)$ on $\hat{X}_W$, we prove the existence of a Fontaine-Lafaille module structure on $(\hat{E}_W, \hat{\nabla}_W)$ (Esnault and Groechenig, Acta Math 225(1):103–158, 2020, Theorem 1.6, Section 4), which, via Faltings' functor, eventually yields a *crystalline* $\mathbb{Z}_{p^f}$*-local system* on the algebraic scheme X_K, where f is the period of the Higgs-de Rham flow. This in turn implies that the rigid complex local systems on $X_{\mathbb{C}}$, for $p > 0$ large so they are integral by p, the residual characteristic of such a good W, descend to crystalline $\mathbb{Z}_{p^f}$-local systems, see Esnault and Groechenig (Acta Math 225(1):103–158, 2020, Section 5). This property remains true even if X is only quasi-projective under a strong cohomological rigidity assumption, which is fulfilled on Shimura varieties of real rank ≥ 2, and assuming in addition that the local monodromies at infinity are unipotent, see Esnault and Groechenig (Frobenius structures and unipotent monodromy at infinity, Preprint 2021, 8 pp., Theorem A.4, Theorem A.22).

9.1 The Main Theorems

We summarize the theorems in the projective case and then in the quasi-projective case separately as the assumptions in the latter case are more technical.

9.1.1 Good Model in the Projective Case

We use the notation of Chap. 8, Sect. 8.5. We denote by $\mathcal{M}$ the rank one Higgs bundle which corresponds to the torsion rank one connection $\mathcal{L}$. Explicitly, if $\mathcal{L} =$

H. Esnault, *Local Systems in Algebraic-Arithmetic Geometry*, Lecture Notes in Mathematics 2337, https://doi.org/10.1007/978-3-031-40840-3_9

(L, ∇) then $\mathcal{M} = (L, 0)$ as $\mathcal{L}$ is assumed to be torsion. Recall from Simpson's theory [Sim92, Lemma 4.5] that on the Dolbeault moduli space $M_{Dol}(X, r, \mathcal{M})$ of stable points over $\mathbb{C}$, we have the $\mathbb{C}^\times$-operation which acts as homotheties on the Higgs field. If (V_i, θ_i), $i = 1, \dots, N$ are the N Higgs bundles associated to the N rigid connections (E_i, ∇_i), $i = 1, \dots, N$, then (V_i, θ_i) is stable under $\mathbb{C}^\times$, thus (E_i, ∇_i) is a polarized complex variation of Hodge structure (E_i, Fil_i, ∇_i) where $Fil_i \subset E_i$ is a locally split filtration which satisfies Griffiths transversality, and

$$(V_i, \theta_i) = (gr(E_i), gr(\nabla_i)),$$

see [Sim92, Lemma 4.5].

With reference to the conditions for the base S of a *good model* in Chap. 8, Sect. 8.5, we request now S to fulfil

(1a) the same as (1) together with the existence of an S-point of X_S;
(2a) the same as (2) for $M_{dR}(X, r, \mathcal{L})$ and $M_{Dol}(X, r, \mathcal{M})$;
(3a) the same as (3) for $M_{dR}(X, r, \mathcal{L})$ and $M_{Dol}(X, r, \mathcal{M})$;
(4a) the same as (4) for $M_{dR}(X, r, \mathcal{L})$ and $M_{Dol}(X, r, \mathcal{M})$;
(5a) the same as (5) for $M_{dR}(X, r, \mathcal{L})$ and $M_{Dol}(X, r, \mathcal{M})$;
(6a) the filtrations $Fil_i \subset E_i$ have a model $Fil_{i,S} \subset E_{iS}$ for $i = 1, \dots, N$, which is locally split;
(7a) for all closed points $s \in |S|$, the characteristic of the residue field of s is $> r+1$ and $> \dim(X)$;
(8a) the local systems $\mathbb{L}_i, i = 1, \dots, N$ associated by the Riemann-Hilbert correspondence to $(E_i, \nabla_i)_{\mathbb{C}}, i = 1, \dots, N$, are integral at all residue characteristics p of the closed points of S.

The condition (6a) can be fulfilled as there are finitely many filtrations Fil_i. The conditions (2a)(3a)(4a)(5a) can be fulfilled applying the existence of the flat moduli spaces $M_{dR}(X, r, \mathcal{L})_S$ and $M_{Dol}(X, r, \mathcal{M})_S$ over S, due to Langer [Lan14, Theorem 1.1] and base change. Note that condition (8a) implies that the $\mathbb{L}_i$ come from étale $\bar{\mathbb{Z}}_p$- local systems on $X_{\mathbb{C}}$, which by base change, is the same as étale $\bar{\mathbb{Z}}_p$- local systems on $X_{\bar{K}}$, where $K = \mathrm{Frac}(W(s))$ and $\mathrm{Spec}(W(s)) \to S$ is a Witt-vector point with residual closed point $\dot{s} \in |S|$. In addition they are irreducible over $\bar{\mathbb{Q}}_p$.

9.1.2 Theorem in the Projective Case

Theorem 9.1 (See [EG20], Section 5) *For any closed point $s \in |S|$ in the basis of a good model, the p-adic local systems $\mathbb{L}_i, i = 1, \dots, N$ on $X_{\bar{K}}$ descend to crystalline p-adic local systems on X_K.*

9.1.3 Good Model in the Quasi-projective Case

We use the notation of Chap. 8, Sect. 8.5 and of 9.1.1. We fix a good compactification $X \hookrightarrow \bar{X}$ with smooth $\bar{X}$ projective such that $D := \bar{X} \setminus X$ is a strict normal crossings divisor. The de Rham moduli space in rank r and fixed determinant $\mathcal{L}$ in the projective case is replaced by the de Rham moduli $M_{dR}(\bar{X}, r, \mathcal{L}, D)$ of connections (E, ∇) on $\bar{X}$ with log-poles along D and with nilpotent residues. We remark, even if we do not use it, that then $\mathcal{L}$ extends to a torsion rank one connection on $\bar{X}$. The nilpotency of the residues implies that the (Betti or de Rham) Chern classes of the underlying bundles E are 0, see [EV86, Proposition B.1]. We have N complex points (E_i, ∇_i) of $M_{dR}(\bar{X}, r, \mathcal{L}, D)$ which describe the 0-dimensional components (rigid objects). Furthermore, [Sim92, Lemma 4.5] is replaced by [Moc06, Theorem 10.5] which guarantees that we have a locally split filtration $Fil_i \subset E_i$ as in the projective case which satisfied Griffiths transversality. Finally

$$(V_i, \theta_i) = (gr(E_i), gr(\nabla_i)),$$

is a stable Higgs bundle with logarithmic poles along D and nilpotent residues and determinant $\mathcal{M} = (L, 0)$. With reference to the conditions for the base S of a *good model* in Chap. 8, Sect. 8.5, and the ones in the projective case, we request now S to fulfil

(1b) $X_S \hookrightarrow \bar{X}_S$ is a model of $X \hookrightarrow \bar{X}$ such that $\bar{X}_S/S$ is smooth projective and $D_S := \bar{X}_S \setminus X_S$ is a relative normal crossings divisor, and there is an S-point of X_S;
(2b) the same as (2) for $M_{dR}(\bar{X}, r, \mathcal{L}, D)$ and $M_{Dol}(\bar{X}, r, \mathcal{M}, D)$;
(3b) the same as (3) for $M_{dR}(\bar{X}, r, \mathcal{L}, D)$ and $M_{Dol}(\bar{X}, r, \mathcal{M}, D)$;
(4b) the same as (4) for $M_{dR}(\bar{X}, r, \mathcal{L}, D)$ and $M_{Dol}(\bar{X}, r, \mathcal{M}, D)$;
(5b) the same as (5) for $M_{dR}(\bar{X}, r, \mathcal{L}, D)$ and $M_{Dol}(\bar{X}, r, \mathcal{M}, D)$;
(6b) the same as (6a) for the Mochizuki filtrations;
(7b) for all closed points $s \in |S|$, the characteristic of the residue field of s in $> 2(r+1)$ and $> \dim(X)$.

The condition (6b) can be fulfilled for the same reason as for 6a): they are finitely many filtrations Fil_i. The conditions (2b)(3b)(4b)(5b) can be fulfilled applying the existence of the flat moduli spaces $M_{dR}(\bar{X}, r, \mathcal{L}, D)_S$ and $M_{Dol}(\bar{X}, r, \mathcal{M}, D)_S$ over S, due to Langer [Lan14, Theorem 1.1] and base change.

9.1.4 Theorem in the Quasi-projective Case

A flat connection (E, ∇) with logarithmitic poles and nilpotent residues is in particular Deligne's extension of its restriction to X, so

$$H^j(X, (E, \nabla)|_X) = H^j(\bar{X}, (\Omega^\bullet_{\bar{X}}(\log D) \otimes_{\mathcal{O}_{\bar{X}}} E, \nabla)).$$

Note that the restriction map

$$H^1(\bar{X}, j_{!*}\mathcal{E}nd^0(E, \nabla)) \hookrightarrow H^j(X, \mathcal{E}nd^0(E, \nabla)|_X)$$

where $\mathcal{E}nd^0$ denotes the trace free endomorphisms, is injective. In particular, if the right hand side vanishes, so does the left hand side. We can then apply the integrality theorem [EG18, Theorem 1.1], see Chap. 7, so all prime numbers are integral for $\mathbb{L}_i, i = 1, \ldots, N$ on X. This explains why the condition which is the counterpart to 8a) is automatically fulfilled under this cohomological assumption and does not appear in the list of conditions.

Theorem 9.2 (See [EG21], Theorem A.22) *Assume that* $H^1(X, \mathcal{E}nd^0(E_i, \nabla_i)) = 0$ *for* $i = 1, \ldots, N$. *Then for all closed points* $s \in |S|$ *of a good model, and all Witt vector points* $\mathrm{Spec}(W(s)) \to S$, *the p-adic local systems* $\mathbb{L}_i, i = 1, \ldots, N$ *on* $X_{\bar{K}}$ *descend to a crystalline p-adic local systems on* X_K *where p is the residue characteristic of s.*

Theorem 9.2 applies for Shimura varieties of real rank ≥ 2. This the framework in which it is applied in the proof of the André-Oort conjecture in [PST21]. The aim of the rest of the Chapter is to formulate the main steps of the proof in the projective case.

9.2 Simpson's Versus Ogus-Vologodsky's Correspondences in the Projective Case

Recall first the Ogus-Vologodsky correspondence [OV07] is a vast elaboration of Deligne-Illusie's splitting of the de Rham complex under the condition that X smooth over a perfect field k lifts to $W_2(k)$ and has dimension $< p$ ([DI87]): for example assume (E, ∇) has nilpotent p-curvature of level one, which means there is an exact sequence $0 \to (F^*S, \nabla_{can}) \to (E, \nabla) \to (F^*Q, \nabla_{can}) \to 0$ where S, Q are coherent sheaves on the Frobenius twist X' of X, $F : X \to X'$ is the relative Frobenius, and ∇_{can} is the canonical connection determined by its flat sections S, Q on X'. Assume (S, Q) are vector bundles. This defines a class in $H^1_{dR}(X, (F^*(Q^{-1} \otimes S), \nabla_{can}))$ which by [DI87, Theorem 2.1] is equal to $H^1(X', Q^{-1} \otimes S) \oplus H^0(X', \Omega^1_{X'} \otimes Q^{-1} \otimes S)$. The class in $H^1(X', Q^{-1} \otimes S)$ yields a vector bundle extension

$$0 \to S \to V' \to Q \to 0$$

on X' and the class in $H^0(X', \Omega^1_{X'} \otimes Q^{-1} \otimes S)$ yields a nilpotent Higgs field

$$\theta' : V' \to Q \to \Omega^1_{X'} \otimes S \to \Omega^1_{X'} \otimes V'.$$

So here we need $p > \dim(X)$ to apply [DI87] *loc. cit.*. Ogus-Vologodksy correspondence assigns $C^{-1}(V', \theta') := (E, \nabla)$ to (V', θ') under the assumption 7a). Assume now that $X = X_s$ for a closed point $s \in |S|$ as in Theorem 9.1. Then starting with $(V_i, \theta_i) = (gr(E_i), gr(\nabla_i))$ on $X_{\mathbb{C}}$, we can restrict (V_i, θ_i) to $(V_i, \theta_i)_s$ on X_s, then take the pull-back $(V_i', \theta_i')_s$ under the arithmetic Frobenius on Spec$k$ where $k = \kappa(s)$ is the residue field of s, and then finally $C_s^{-1}(V_i', \theta_i')_s$. What is its relation to the restriction $(E_i, \nabla_i)_s$ of (E_i, ∇_i)? Here C_s is the Cartier-Ogus-Vologodsky functor on X_s. The miracle is the following theorem.

Theorem 9.3 ([EG20], Proposition 3.5) *$C_s^{-1}(V_i', \theta_i')_s$ is equal to one of the $(E_u, \nabla_u)_s$ where $u \in \{1, \ldots, N\}$.*

Ideal of Proof As C_s^{-1} preserves the stability, see [Lan14, Corollary 5.10], the key point is the so-called Beauville-Narasimhan-Ramanan correspondence [EG20, Theorem 2.17] as proved by Michael Groechenig in his PhD Thesis [Gro16, Proposition 3.15]. In a sense it yields the Ogus-Vologodsky correspondence at the level of moduli spaces (on a curve) with a *scheme theoretic structure*. If $C_s^{-1}(V_i', \theta_i')$ was not rigid, there would be a deformation of its moduli point yielding a nilpotent structure around this point of order as high as we want. This nilpotent structure, transported on the side of the Dolbeault moduli space via the Ogus-Vologodsky correspondence, yields by rigidity over $\mathbb{C}$ and the property 5a) a contradiction, as over $\mathbb{C}$ in $M_{dR}(X, r, \mathcal{L})$ the multiplicities of the isolated points are given, thus bounded on S. □

Remark 9.4 The proof above unlocked the program we had in order to prove the arithmetic crystalline properties associated to rigid connections in Theorem 9.1. We relagate to [EG20, Corollary A.7] a proof based on the same principle showing nilpotency of the p-curvature (which we discussed via p-adic methods in Theorem 8.4), which, in the words of an anonymous referee, yields a canonical action of $\mathbb{G}_m^{\#}$ on the *category* of flat connections on X_s, where $\mathbb{G}_m^{\#}$, is the PD hull of the neutral element of $\mathbb{G}_m$.

9.3 Periodic de Rham-Higgs Flow on X_s and the $GL_r(\bar{\mathbb{F}}_p)$-local Systems on X_K in the Projective Case

As C_s^{-1} is an equivalence of categories (when it is defined), we see that in the proof of Theorem 9.3 the assignment $i \mapsto u$ is a bijection of $\{1, \ldots, N\}$.

Defining a *de Rham-Higgs flow* as a sequence on X_s of $\{(E_\iota, Fil_\iota, \nabla_\iota)\}$ with

$$(E_{\iota+1}, \nabla_{\iota+1}) = C_s^{-1}(grE_\iota, gr\nabla_\iota),$$

we see that for any (E_ι, ∇_ι) with $\iota = 1 \ldots, N$, $(E_\iota, Fil_\iota, \nabla_\iota)$ defines a *periodic* de Rham-Higgs flow of period $f(\iota)|N!$.

By [OV07, Subsection 4.6], refined in [LSZ19, Corollary 3.10] to take into account the possibility of $f(\iota) > 1$, this defines a Fontaine-Lafaille module on X_s, which we do not define here as our emphasis is on the local system, see [FL82, Theorem 3.3]. By the enhancement of the Fontaine-Lafaille construction in *loc. cit.* due to Faltings [Theorem 2.6*][Fal88] (see [EG20, Proposition 4.3]), we obtain via Fontaine-Lafaille-Faltings' functor the following claim.

Claim 9.5 For a closed point s of a good model S, we assign pairwise non-isomorphic local systems $\mathcal{L}_i(p)$ with values in $GL_r(\mathbb{F}_{p^{f(i)}})$ on X_K, for $i = 1, \ldots, N$.

Remark 9.6 The second miracle is provided by the theory of Fontaine-Lafaille-Faltings here: the local systems are defined over the *algebraic* X_K.

9.4 Periodic de Rham-Higgs Flow on $\hat{X}_W$ and the $GL_r(W(\bar{\mathbb{F}}_p))$-local Systems on X_K in the Projective Case

The functor C_s^{-1} of Ogus-Vologosdsky extends to C_n^{-1} on $X_W \otimes_W W_n$ for all n, defining $\hat{C}$ on $\hat{X}_W$. The notion of de Rham-Higgs flow on the formal model $\hat{X}_W$ is completely similar to the one on X_s. See [LSZ19, Corollary 3.10, Theorem 4.1], [SYZ22, 1.2.1] and the work of Xu [Xu19] for very related methods and results. Then:

Claim 9.7 For any $(\hat{E}_{\iota,W}, \hat{\nabla}_{\iota,W})$, the restriction of (E_ι, ∇_ι) to $\hat{X}_W$, with $\iota = 1 \ldots, N$, the triple $(E_\iota, Fil_\iota, \nabla_\iota)$, defines a *periodic* de Rham-Higgs flow of period $f(\iota)|N!$.

By [OV07, Subsection 4.6], refined in [LSZ19, Corollary 3.10] to take into account the possibility of $f(\iota) > 1$, this defines a Fontaine-Lafaille module on $\hat{X}_W$, see [FL82, Theorem 3.3]. By the enhancement of the Fontaine-Lafaille construction in *loc. cit.* due to Faltings [Theorem 2.6*][Fal88] (see [EG20, Proposition 4.3]), we obtain via the Fontaine-Lafaille-Faltings functor the following claim.

Claim 9.8 For any closed point s of a good model S, we assign pairwise non-isomorphic p-adic local systems $\mathcal{L}_i(W)$ with values in $GL_r(W(\mathbb{F}_{p^{f(i)}}))$ on X_K, for $i = 1, \ldots, N$, which are crystalline.

Remark 9.9 This refinement of Faltings concerning crystallinity (also in the quasi-projective case) is precisely the important property used in [PST21].

9.5 From Crystalline p-Adic Local Systems on X_K to p-Adic Local Systems on $X_{\bar{K}}$ in the Projective Case

The local systems $(\mathbb{L}_i, i = 1, \ldots, N)$ are integral at p, thus by the identification $\pi_1(X_{\mathbb{C}}, x_{\mathbb{C}})$ with $\pi_1(X_{\bar{K}}, x_{\bar{K}})$, they become p-adic local systems on $X_{\bar{X}}$. We also have the base change

$$\mathcal{L}_i(W)|_{X_{\bar{K}}} =: \mathcal{L}_i(\bar{K})$$

to $X_{\bar{K}}$. To conclude Theorem 9.1 we should make sure that the $\mathcal{L}_i(\bar{K})$ are not identified on $X_{\bar{K}}$ and really come from the topological rigid local systems.

Theorem 9.10 *The set of p-adic local systems* $\{\mathcal{L}_1(\bar{K}), \ldots, \mathcal{L}_N(\bar{K})\}$ *is, up to order, the set* $\{\mathbb{L}_1, \ldots, \mathbb{L}_N\}$, *viewed as p-adic local systems on* $X_{\bar{K}}$. *Expressed the other way around, the p-adic local systems* $\mathbb{L}_i$ *on* $X_{\bar{K}}$ *descend to crystalline p-adic local systems on* X_K, *for* $i = 1, \ldots, N$ *which in addition have values in* $GL_r(W(\bar{\mathbb{F}}_p))$.

Proof *(Idea of Proof)* The construction of the Fontaine-Lafaille modules over $\hat{X}_W$ is based on the fact that $\hat{X}_W$ is a non-ramified lift of X_s. We could repeat the construction on $\hat{X}_{W(\bar{s})}$ so as to obtain in this way local systems on $X_{\mathrm{Frac}(W(\bar{\mathbb{F}}_p))}$. However, we have to go all the way up to $X_{\bar{K}}$.

To perform this, we use in [EG20, Section 5] Faltings' p-adic Simpson correspondence. In order to be able to apply Faltings' theory, we have to be sure that the Higgs bundles on $X_{\bar{K}}$ and the $\mathcal{L}_i(\bar{K})$ are small in his sense. We can find infinitely many prime numbers $p > 0$ with the property that $\kappa(s) = \mathbb{F}_p$ and all the $f_i = 1$. This is the content of [EG20, Lemma 5.11] and yields a weaker form of Theorem 9.1. We do not get the result on all closed points of S but on an infinity of those with infinite different residual characteristics.

In general, we resort to [SYZ22], which relies on Faltings' p-adic Simpson correspondence, and "does" part of the Fontaine-Lafaille-Faltings program in the ramified case. We first argue that $\mathcal{L}_i(\bar{K}) \otimes_{\bar{Z}_p} \bar{\mathbb{F}}_p$ is irreducible. If not, using the W-point in 1a) and the consequence that $\pi_1(X_K)$ is then a semi-direct product of $\mathrm{Gal}(\bar{K}/K)$ with $\pi_1(X_{\bar{K}})$, we can kill on $\mathrm{Gal}(\bar{K}/K)$ the $GL_r(\bar{\mathbb{F}}_p)$-representation by a finite field extension K'/K. This enables us to conclude that $\mathcal{L}_i(K')$ itself is reducible, which by [SYZ22, Theorem 5.15] violates the stability of the associated Higgs bundle on $X_{K'}$. We argue similarly to distinguish the $\mathcal{L}_i(\bar{K}) \otimes_{\bar{Z}_p} \bar{\mathbb{F}}_p$ for $i = 1, \ldots, N$. This finishes the proof. □

Remark 9.11 We hope in the near future ([EG22]) to strengthen the results and shorten the proofs of the existing ones using more and in particular more recent p-adic methods.

9.6 Remarks

Simpson's non-abelian Hodge theory [Sim92] is the starting point of this chapter. We also mention that Deligne's mixed Hodge theory [Del71] yields a rational mixed Hodge structure on the Malčev completion of the topological fundamental group, see [Mor78, Theorem 9.2]. One more recent way to see this is to use that the truncations of the Malčev completion are described as the cohomology groups of a simplicial scheme (Beilinson, see [DG05, Proposition 3.4]). There are ℓ-adic variants of it, see [Pri09].

Chapter 10
Lecture 10: Comments and Questions

Abstract The aim of this last Chapter is to list a few questions encountered during the Lectures.

10.1 With Respect to the p-Curvature Conjecture (Chap. 2)

What would be a formulation of the p-curvature conjecture when we replace a smooth quasi-projective variety over $\mathbb{C}$ by the formal completion along a non-normal subvariety of a smooth variety over $\mathbb{C}$? The question is motivated by a discussion with Johan de Jong on [LR96].

10.2 With Respect to the Malčev-Grothendieck's Theorem and Its Shadows in Characteristic $p > 0$ (Chap. 3)

As already mentioned in Sect. 3.3 we have two versions of Malčev-Grothendieck theorem over an algebraically closed field of characteristic $p > 0$, one on the infinitesimal site (Gieseker's conjecture, solved, see Theorem 3.4), one on the crystalline site (unsolved) due to de Jong, but we do not have at present a formulation in the prismatic site, one difficulty being the *iso*-notion.

10.3 With Respect to Lubotzky's Theorem 4.2

Notation as in *loc. cit.*. Assume π is a profinite group. What can be said on

$$\dim_{\mathbb{F}_\ell} H^i(\pi, \rho) \text{ for any } i \in \mathbb{N}$$

where $\rho : \pi \to GL_r(\mathbb{F}_\ell)$ is a linear representation.

H. Esnault, *Local Systems in Algebraic-Arithmetic Geometry*, Lecture Notes in Mathematics 2337, https://doi.org/10.1007/978-3-031-40840-3_10

It holds $\dim_{\mathbb{F}_\ell} H^0(\pi, \rho) \leq r$ for any representation, whatever π is.

It holds $\dim_{\mathbb{F}_\ell} H^1(\pi, \rho) \leq \delta \cdot r$ for any profinite group π which is topologically generated by δ elements, as a 1-continuous cocycle is uniquely defined by its value on topological generators of π.

We observe that this linear bound for H^1 is not equivalent to the finite generation of the profinite group π. Here is an example due to Lubotzsky. Let G be a finite non-abelian simple group, let $\pi = \prod_0^\infty G$ be the infinite product of G with itself, endowed with the profinite structure stemming from all finite quotients. As a finite quotient $\pi \twoheadrightarrow Q$ has to factor through some G^m for some $m \in \mathbb{N}$, Q itself has to be isomorphic to some G^n for some $n \in \mathbb{N}$. As $\dim_{\mathbb{F}_\ell} H^1(\pi, \rho) \leq \dim_{\mathbb{F}_\ell} H^1(\pi, \rho^{ss})$ where ρ^{ss} is the semi-simplification of ρ, we may assume that ρ is irreducible. We write the exact sequence $1 \to \mathrm{Ker}(\rho) \to \pi \to \rho(\pi) \cong G^n \to 1$. Then $\mathrm{Ker}(\rho)$, which is abstractly isomorphic to π, has no abelian quotient. Thus $H^1(\mathrm{Ker}(\rho), \rho|_{\mathrm{Ker}(\rho)}) = 0$ and the Hochshild-Serre spectral sequence yields an isomorphism $H^1(\rho(\pi), \rho) \xrightarrow{\cong} H^1(\pi, \rho)$. By [AG84, Theorem A], $\dim_{\mathbb{F}_\ell} H^1(\rho(\pi), \rho) \leq r$. This finishes the proof.

For $i = 2$ Lubotzky's theorem 4.2 yields even a characterization of finite presentation.

What does the growth of the cohomology for $i \geq 3$ encode as a property?

Do we have special properties for $\dim_{\mathbb{F}_\ell} H^i(\pi, \rho)$ for any $i \geq 3$ if π is the tame fundamental group of X smooth quasi-projective over any algebraically closed field?

10.4 With Respect to Theorem 4.7

Notation as in *loc. cit.*. It is wishful that Theorem 4.5 be true under the assumption that X is quasi-projective normal over an algebraically closed characteristic $p > 0$ field k.

If $j : X \hookrightarrow \bar{X}$ is a normal compactification of a normal X over k, is there a formula which enables one to bound $H^2(\pi_1^t(X), M)$ by some cohomological invariant of a constructible sheaf on $\bar{X}$ built out of the local system $\underline{M}$?

What about higher cohomology $H^i(\pi_1^t(X), M)$, $i \geq 3$, also for X normal (see Sect. 10.3)?

10.5 With Respect to Theorems 5.3 and 5.10

Can we extend those theorems to formally smooth proper schemes, to smooth rigid spaces, is there a version for normal varieties, for non-normal varieties involving the pro-étale fundamental group developed in [Sch13] and in all generality in [BS15] etc. We can also raise similar questions concerning the finite presentation of the (tame) fundamental group. (The question for formal and rigid spaces was posed to us by Piotr Achinger and Ben Heurer.)

10.6 With Respect to Theorem 6.2

Can we replace in the formulation of Theorem 6.2 the Betti moduli in a given rank of a normal complex variety by the Mazur or Chenevier deformation space of a given $\bar{\mathbb{F}}_\ell$-residual representation over X over $\bar{\mathbb{F}}_p$ to obtain in those spaces Zariski density of the local systems with quasi-unipotent monodromies at infinity?

10.7 With Respect to Theorem 7.6 and a More Elaborate Version of Theorem 7.8, See Theorem [dJE22, Theorem 1.4]

The main question is "where" is $sp^{-1}_{\mathbb{C},\bar{s}}(\mathbb{L}^\sigma_\ell)$ located with respect to $\mathbb{L}_\ell$ on $M_B(X, r, \mathcal{L}, T_i)$. If $\mathbb{L}_\ell$ lied on a d-dimensional component, what about $sp^{-1}_{\mathbb{C},\bar{s}}(\mathbb{L}^\sigma_\ell)$? For example, if $X = \mathbb{P}^1 \setminus \Sigma$ where $\Sigma \subset \mathbb{P}^1$ consists of finitely many points, then irreducible rigid local systems are cohomologically rigid ([Kat96, Corollary 1.2.5]), thus the proof of Theorem 7.6 shows that then if $d = 0$, then $sp^{-1}_{\mathbb{C},\bar{s}}(\mathbb{L}^\sigma_\ell)$ is a 0-dimensional component as well. What about the higher dimensional components in this case?

10.8 With Respect to Theorem 7.8

Does Theorem 7.8 hold with X defined over $\bar{\mathbb{F}}_p$? To be precise, assume that in a given rank r, there is an irreducible ℓ-adic local system. Can we conclude that for all $\ell' \neq p$ there is an irreducible ℓ'-adic local system of rank r, and that perhaps by $\ell' = p$, there is an irreducible isocrystal of rank r? Note if arithmetic ℓ-adic local systems were dense in the Chenevier deformation space as wished for in Conjectures [EK22, Weak and Strong Conjectures], this much weaker problem would have a positive answer.

10.9 With Respect to Chaps. 8 and 9

As the de Rham-crystalline side of the theory does not request any cohomological condition on the rigid systems when X is projective, if would be wishful to understand the full strength of the theorems in the non-proper case.

Reference

[AE19] T. Abe, H. Esnault, A Lefschetz theorem for overconvergent isocrystals with Frobenius structure. Ann. Ecole Norm. Sup. **52**(4), 1243–1264 (2019)

[Ach17] P. Achinger, Wild ramification and $K(\pi, 1)$ spaces. Invent. Math. **210**(2), 453–499 (2017)

[AZ17] P. Achinger, M. Zdanowicz, Some elementary examples of non-liftable varieties, Proc. Am. Math. Soc. **145**, 4717–4729 (2017)

[And74] M. Anderson, Exactness properties of profinite completion functors. Topology **13**, 229–239 (1974)

[And89] André, Y., *G-Functions and Geometry*. Aspects of Mathematics, vol. E13 (Vieweg Verlag, Braunschweig, 1989)

[And04] Y. André, Sur la conjecture des p-courbures de Grothendieck-Katz et un problème de Dwork, in *Geometric Aspects of Dwork Theory*, vol. I, II (Walter de Gruyter, Berlin, 2004), pp. 55–112

[Ara01] A. Arabia, Relèvements des algèbres lisses et de leurs morphismes. Comment. Math. Helv. **76**(4), 607–639 (2001)

[AS16] J. Aramayona, J. Souto, Rigidity phenomena in the mapping class group, in *Handbook of Teichmüller Theory*, vol. 27 (European Mathematical Society, 2016), pp. 131–165

[AG84] M. Aschbacher, M. Guralnick, Some applications of the first cohomology group. J. Algebra **90**, 446–460 (1984)

[BGH18] C. Barwick, S. Glasman, P. Haine, *Exodromy*. https://arxiv.org/pdf/1807.03281.pdf

[Bau04] M. Bauer, Über einen Satz von Kronecker. Arch. d. Math. u. Phys. **6**, 212–222 (1904)

[BBV23] O. Becker, E. Breuillard, P. Varjú, Random character varieties (2022, in progress)

[BBDG82] A. Beilinson, J. Bernstein, P. Deligne, Faisceaux pervers, in *Analysis and Topology on Singular Spaces, I* (Luminy 1981). Astérisque, vol. 100 (Société mathématique de France, Paris, 1982), pp. 5–171

[Ber74] P. Berthelot, *Cohomologie cristalline des schémas de caractéristique* $p > 0$. Lecture Notes in Mathematics, vol. 407 (Springer, Berlin, 1974), 604 pp.

[Ber00] P. Berthelot, $\mathcal{D}$-modules arithmétiques II. Descente par Frobenius. Mém. Soc. Math. Fr. **81**, vi+136 pp. (2000)

[BL17] B. Bhatt, J. Lurie, A Riemann-Hilbert correspondence in characteristic $p > 0$. https://www.math.ias.edu/~lurie/papers/ModpRH.pdf

[BS15] B. Bhatt, P. Scholze, The pro-étale topology for schemes. Astérisque **369**, 99–201 (2015)

[BS21] B. Bhatt, P. Scholze, Prismatic F-crystals and crystalline Galois representations. http://www.math.uni-bonn.de/people/scholze/PrismaticCrystals.pdf

H. Esnault, *Local Systems in Algebraic-Arithmetic Geometry*, Lecture Notes in Mathematics 2337, https://doi.org/10.1007/978-3-031-40840-3

[Bel80] G. Belyĭ, Galois extensions of a maximal cyclotomic field, translated by Neal Koblitz. Math. USSR Izv. **14**(2), 247–256 (1980)

[BGMW22] I. Biswas, S. Gupta, M. Mj, J.P. Whang, Surface group representations in $SL2(\mathbb{C})$ with finite mapping class orbits. Geom. Topol. **26**(2), 679–719 (2022)

[BK06] G. Böckle, C. Khare, Mod ℓ representations of arithmetic fundamental groups. II. A conjecture of A. J. de Jong. Compos. Math. **142**(2), 271–294 (2006)

[Bos01] J.-B. Bost, Algebraic leaves of algebraic foliations over number fields. Publ. Math. I. H. É. S. **93**, 161–221 (2001)

[BKT13] J. Brunebarbe, B. Klingler, B. Totaro, Symmetric differentials and the fundamental group. Duke Math. J. **162**, 1–17 (2013)

[Che14] G. Chenevier, The p-adic analytic space of pseudocharacters of a profinite group and pseudorepresentations over arbitrary rings, in *Automorphic Forms and Galois Representations*, vol. 1 (2011), pp. 221–285. Lecture Note Series, vol. 414 (London Mathematical Society, 2014)

[Chu85] D. Chudnovsky, G. Chudnovsky, Applications of Padé approximation to the Grothendieck conjecture on linear differential equations, in *Number Theory*. Lecture Notes in Mathematics, vol. 1135 (Springer Verlag, 1985), pp. 52–100

[D'Ad20] M. D'Addezio, Some remarks on the companions conjecture for normal varieties. https://arxiv.org/abs/2006.09954

[dJ01] J. de Jong, A conjecture on the arithmetic fundamental group. Israel J. Math. **121**, 61–84 (2001)

[dJEG22] J. de Jong, H. Esnault, M. Groechenig, Rigid non-cohomologically rigid local systems, in *Algebraic Geometry and Physics*. Preprint 2022, 6 pp., vol. 1 (2023, to appear). http://page.mi.fu-berlin.de/esnault/helene_publ.html

[dJE22] J. de Jong, H. Esnault, Integrality of the Betti moduli space. Preprint 2022, 18 pp. Trans. AMS (to appear). https://page.mi.fu-berlin.de/esnault/preprints/helene/146_dJ_esn.pdf

[Del70] P. Deligne, Équations différentielles à points singuliers réguliers. Lecture Notes in Mathematics, vol. 163 (Springer, Berlin, 1970)

[Del71] P. Deligne, Théorie de Hodge: II. Publ. math. I.H.É.S. **40**, 5–57 (1971)

[Del72] P. Deligne, Les constantes des équations fonctionnelles des fonctions L, in *Proc. Antwerpen Conference*, vol. 2. Lecture Notes in Mathematics, vol. 349 (Springer, Berlin, 1972), pp. 501–597

[Del73] P. Deligne, Comparaison avec la théorie transcendente, in *SGA 7 II, Exposé XIV*. Lecture Notes in Mathematics, vol. 340 (1973), pp. 116–164

[Del80] P. Deligne, La conjecture de Weil II. Publ. math. I.H.É.S. **52**, 137–252 (1980)

[DI87] P. Deligne, L. Illusie, it Relèvements modulo p^2 et décomposition du complexe de de Rham. Invent. Math. **89**(2), 247–270 (1987)

[DG05] P. Deligne, A. Goncharov, Groupes fondamentaux motiviques de Tate mixte. Ann. Sci. Éc. Norm. Sup. 4o série **38**, 1–56 (2005)

[Del12] P. Deligne, Finitude de l'extension de $\mathbb{Q}$ engendrée par des traces de Frobenius, en caractéristique finie. Mosc. Math. J. **12**(3), 497–514 (2012)

[Dri01] V. Drinfeld, On a conjecture of Kashiwara. Math. Res. Lett. **8**(5–6), 713–728 (2001)

[Dri12] V. Drinfeld, On a conjecture of Deligne. Volume dedicated to the memory of I. M. Gelfand. Moscow Math. J. **12**(3), 515–542 (2012)

[Esn17] H. Esnault, Survey on some aspects of Lefschetz theorems in algebraic geometry. Rev. Mat. Complut. **30**(2), 217–232 (2017)

[EM10] H. Esnault, V. Mehta, Simply connected projective manifolds in characteristic $p > 0$ have no nontrivial stratified bundles. Invent. Math. **181**, 449–465 (2010)

[EL13] H. Esnault, A. Langer, On a positive equicharacteristic variant of the p-curvature conjecture. Doc. Math. **18**, 23–50 (2013)

[ES15] H. Esnault, A. Shiho, Existence of locally free lattices of crystals. Preprint 2015, 2 pp. https://page.mi.fu-berlin.de/esnault/preprints/helene/119b_esn_shi.pdf

[ES18] H. Esnault, A. Shiho, Convergent isocrystals on simply connected varieties. Ann. Inst. Fourier **68**(5), 2019–2148 (2018)

[ES19] H. Esnault, A. Shiho, Chern classes of crystals. Trans. AMS **371**(2), 1333–1358 (2019)

[EK18] H. Esnault, M. Kisin, D-modules and finite monodromy. Sel. Math. Volume dedicated to Alexander Beilinson **24**(1), 217–232 (2018)

[EG18] H. Esnault, M. Groechenig, Cohomologically rigid connections and integrality. Sel. Math. **24**(5), 4279–4292 (2018)

[EG20] H. Esnault, M. Groechenig, Rigid connections and F-isocrystals. Acta Math. **225**(1), 103–158 (2020)

[EG21] H. Esnault, M. Groechenig, Frobenius structures and unipotent monodromy at infinity. Preprint 2021, 8 pp. http://page.mi.fu-berlin.de/esnault/preprints/helene/143_appPST.pdf

[EG22] H. Esnault, M. Groechenig (in progress)

[EK11] H. Esnault, M. Kerz, Notes on Deligne's letter to Drinfeld dated March 5 (2007) https://page.mi.fu-berlin.de/esnault/preprints/helene/103--110617.pdf

[EK12] H. Esnault, M. Kerz, A finiteness theorem for Galois representations of function fields over finite fields (after Deligne). Acta Math. Vietnam. **37**(4), 531–562 (2012)

[EK20] H. Esnault, M. Kerz, Arithmetic subspaces of moduli spaces of rank one local systems. Camb. J. Math. **8**(3), 453–478 (2020)

[EK21] H. Esnault, M. Kerz, Étale cohomology of rank one ℓ-adic local systems in positive characteristic. Sel. Math. **27**(4), paper 58, 25 pp. (2021)

[EK22] H. Esnault, M. Kerz, Density of arithmetic representations of function fields Epiga **6**, 18 pp. (2022)

[EK23] H. Esnault, M. Kerz, Local systems with quasi-unipotent monodromy at infinity are dense. Preprint 2021, 9 pp. Israel J. Math. (to appear). http://page.mi.fu-berlin.de/esnault/preprints/helene/140_esn_kerz.pdf

[EK16] H. Esnault, L. Kindler Lefschetz theorems for tamely ramified coverings. Proc. AMS **144**, 5071–5080 (2016)

[ESS22] H. Esnault, M. Shusterman, V. Srinivas, Finite presentation of the tame fundamental group. Sel. Math. **28**(2), paper 37, 19 pp. (2022)

[ESS22b] H. Esnault, V. Srinivas, J. Stix, An obstruction to lifting to characteristic 0. Algebr. Geom. **10**(3), 327–347 (2023)

[EV86] H. Esnault, E. Viehweg, Logarithmic de Rham complexes and vanishing theorems. Invent. Math. **86**, 161–194 (1986)

[Fal83] G. Faltings, Endlichkeitssätze für abelsche Varietäten über Zahlkörpern. Invent. Math. **73**(3), 349–366 (1983)

[Fal88] G. Faltings, Crystalline cohomology and p-adic Galois-representations, in *Algebraic Analysis, Geometry, and Number Theory (Baltimore, MD, 1988)* (Johns Hopkins University Press, Baltimore, 1989), pp. 25–80

[FL82] J.M. Fontaine, G. Laffaille, Construction de représentations p-adiques. Ann. Sci. Éc. Norm. Sup. **15**, 547–608 (1982)

[Gai07] D. Gaitsgory, On de Jong's conjecture. Israel J. Math. **157**, 155–191 (2007)

[Gie75] D. Gieseker, Flat vector bundles and the fundamental group in non-zero characteristics. Ann. Sc. Norm. Super. Pisa **4** Sér. 2(1), 1–31 (1975)

[Gri70] P. Griffiths, Periods of integrals on algebraic manifolds: summary of results and discussion of open problems. Bull. Am. Math. Soc. **76**, 228–296 (1970)

[GM88] W. Goldman, J. Millson, The deformation theory of representations of fundamental groups of compact Kähler manifolds. Publ. math. I.H.É.S. **67**, 43–96 (1988)

[Gro16] M. Groechenig, Moduli of flat connections in positive characteristic. Math. Res. Lett. **23**(4), 989–1047 (2016)

[Gro70] A. Grothendieck, Représentations linéaires et compactifications profinies des groupes discrets. Manuscr. Math. **2**, 375–396 (1970)

[Gro57] A. Grothendieck, Sur quelques points d'algèbre homologique. Tohoku Math. J. **9**(2), 119–221 (1957)

[Hru04] E. Hrushovski, The elementary theory of Frobenius automorphisms. http://de.arxiv.org/pdf/math/0406514v1

[Ill79] L. Illusie, Complexe de de Rham-Witt et cohomologie cristalline. Ann. scient. Éc. Norm. Sup. **12**, 501–661 (1979)

[Ill81] L. Illusie, Théorie de Brauer et caractéristique d'Euler-Poincaré d'après P. Deligne. Astérisque **82–83**, 161–172 (1981)

[Kas81] M. Kashiwara, Quasi-unipotent constructible sheaves. J. Fac. Sci. Univ. Tokyo, Sect. IA Math. **28**(3), 757–773 (1981)

[Kat70] N. Katz, Nilpotent connections and the monodromy theorem: applications of a result of Turrittin. Publ. Math. I.H.É.S. **39**, 175–232 (1970)

[Kat72] N. Katz, Algebraic solutions of differential equations (p-curvature and the Hodge filtration). Invent. Math. **18**, 1–118 (1972)

[Kat96] N. Katz, *Rigid Local Systems* (Princeton University Press, Princeton, 1996)

[KS09] M. Kerz, A. Schmidt, Covering data and higher dimensional global class field theory. J. Number Theory **129**(10), 2569–2599 (2009)

[KS10] M. Kerz, A. Schmidt, On different notions of tameness in arithmetic geometry. Math. Ann. **346**(3), 64–668 (2010)

[KM74] N. Katz, W. Messing, Some consequences of the Riemann hypothesis for varieties over finite fields. Invent. Math. **23**, 73–77 (1974)

[Ked22] K. Kedlaya, Étale and crystalline companions. Épiga (to appear). https://kskedlaya.org/papers/

[KS10] M. Kerz, A. Schmidt, On different notions of tameness in arithmetic geometry. Math. Ann. **346**(3), 641–668 (2010)

[KP22] C. Klevdal, S. Patrikis, G-cohomologically rigid local systems are integral. Trans. Am. Math. Soc. **375**(6), 4153–4175 (2022)

[Kro57] L. Kronecker, Zwei Sätze über Gleichungen mit ganzzahligen Coefficienten. J. Reine U. Angew. Math. **53**, 153–175 (1857)

[Laf02] L. Lafforgue, Chtoucas de Drinfeld et correspondance de Langlands. Invent. Math. **147**(1), 1–241 (2002)

[Lam22] Y.H.J. Lam, Motivic local systems on curves and Maeda's conjecture. https://arxiv.org/pdf/2211.06120.pdf

[Lan04] E. Landau, Anwendung des Eisensteinschen Satzes auf die Theorie der Gaussschen Differentialgleichung. J. Reine Angew. Math. (Crelles J.) **127**, 92–102 (1904)

[LSZ19] G. Lan, M. Sheng, K. Zuo, Semistable Higgs bundles, periodic Higgs bundles and representations of algebraic fundamental groups. J. Eur. Math. Soc. **21**(10), 3053–3112 (2019)

[LL22a] A. Landesman, D. Litt, Geometric local systems on very general curves and isomonodromy. https://arxiv.org/pdf/2202.00039.pdf

[LL22b] A. Landesman, D. Litt, Canonical representations of surface groups. https://arxiv.org/abs/2205.15352

[Lan14] A. Langer, Semistable modules over Lie algebroids in positive characteristic. Doc. Math. **19**, 509–540 (2014)

[LR96] B. Lasell, M. Ramachandran, Observations on harmonic maps and singular varieties. Ann. Sci. É.N.S. **29** 4o série, 135–148 (1996)

[LLSS20] W. Li, D. Litt, N. Salter, P. Srinivasan, Surface bundles and the section conjecture. https://arxiv.org/pdf/2010.07331.pdf

[Lub01] A. Lubotzky, Pro-finite Presentations. J. Algebra **242**(2), 672–690 (2001)

[Mal40] A. Malčev, On isomorphic matrix representations of infinite groups. Mat. Sb. N. S. **8**(50), 405–422 (1940)

[Maz89] B. Mazur, Deforming galois representations, in *Galois Groups Over* $\mathbb{Q}$ (Proc. Workshop, Berkeley/CA (USA) (1987), pp. 385–437. Mathematical Sciences Research Institute Publications, vol. 16 (Springer, New York, 1989)

[Moc06] T. Mochizuki, Kobayashi-Hitchin correspondence for tame harmonic bundles and an application. Astérisque **309**, viii+117 (2006)
[Moc07] T. Mochizuki, Asymptotic behaviour of tame harmonic bundles and an application to pure twistor D-modules. II. Mem. Am. Math. Soc. **185**(870), xii+565 pp. (2007)
[Mor78] J. Morgan, The algebraic topology of smooth algebraic varieties. Publ. Math. Inst. Hautes Études Sci. **48**, 137–204 (1978)
[OV07] A. Ogus, V. Vologodksy, Nonabelian Hodge theory in characteristic p. Publ. Math. Inst. Hautes Études Sci. **106**, 1–138 (2007)
[PSW21] A. Patel, A. Shankar, J.P. Whang, The rank two p-curvature conjecture on generic curves. Adv. Math. **386**, 107800, 33 pp. (2021)
[Pet22] A. Petrov, Geometrically irreducible p-adic local systems are de Rham up to a twist. Duke (to appear). https://arxiv.org/abs/2012.13372
[PST21] J. Pila, A. Shankar, J. Tsimerman, Canonical heights on Shimura varieties and the André-Oort conjecture. https://arxiv.org/pdf/2109.08788.pdf
[Pri09] J.P. Pridham, Weight decompositions on étale fundamental groups. Am. J. Math. **131**(3), 869–891 (2009)
[Ray78] M. Raynaud, Contre-exemple au "Vanishing Theorem" en caractéristique $p > 0$, in *C. P. Ramanujam - A Tribute*. Studies in Mathematics, vol. 8 (Tata Institute of Fundamental Research, Bombay, 1978), pp. 273–278
[Roq70] P. Roquette, Abschätzung der Automorphismenanzahl von Funktionenkörpern bei Primzahlcharakteristik. Math. Z. **117**, 157–163 (1970)
[Sch13] P. Scholze, p-adic Hodge theory for rigid-analytic varieties. Forum Math. Pi **1**, e1, 77 pp. (2013)
[Ser61] J.-P. Serre, Exemples de variétés projectives en caractéristique p non relevables en car-actéristique zéro. Proc. Natl. Acad. Sci. USA **47**, 108–109 (1961)
[Sha18] A. Shankar, The p-curvature conjecture and monodromy around simple closed loops. Duke Math. J. **167**(10), 1951–1980 (2018)
[Sim92] C. Simpson, Higgs bundles and local systems. Publ. Math. I.H.É.S. **75**, 5–95 (1992)
[SYZ22] R. Sun, J. Yang, K. Zuo, Projective crystalline representations of étale fundamental groups and twisted periodic Higgs-de Rham flow. J. Eur. Math. Soc. **24**(6), 1991–2076 (2022)
[vDdB21] R. van Dobben de Bruyn, A variety that cannot be dominated by one that lifts. Duke J. Math. **170**(7), 1251–1289 (2021)
[Var18] Y. Varshavski, Intersection of a correspondence with a graph of Frobenius. J. Algebraic Geom. **27**(1), 1–20 (2018)
[Wie06] G. Wiesend, A construction of covers of arithmetic schemes. J. Number Theory **121**(1), 118–131 (2006)
[Xu19] D. Xu, Relèvement de la transformée de Cartier d'Ogus-Vologodsky modulo p^n. Mém. SMF **163**, 6–144 (2019)
[Zuo00] K. Zuo, On the negativity of kernels of Kodaira-Spencer maps on Hodge bundles and applications. Asian J. Math. **4**, 279–302 (2000)
[EGAIV$_4$] Éléments de Géométrie Algébrique Étude locale des schémas et des morphismes de schémas. Quatrième partie. Publ. Math. I.H.É.S. **32**, 5–361 (1967)
[SGA1] Séminaire de Géométrie Algébrique: Revêtements étales et groupe fondamental. Lecture Notes in Mathematics, vol. 224 (Springer, Berlin, 1971)
[SGA4] Séminaire de Géométrie Algébrique: Théorie des Topos et cohomologie étale des schémas, Tome 3. Lecture Notes in Mathematics, vol. 305 (Springer, Berlin, 1973)
[SGA7.2] Séminaire de Géométrie Algébrique: Groupes de monodromie en géométrie algébrique. Lecture Notes in Mathematics, vol. 340 (Springer, Berlin, 1973)
[SP] Stack Project. https://stacks.math.columbia.edu

LECTURE NOTES IN MATHEMATICS 

Editorial Policy

1. Lecture Notes aim to report new developments in all areas of mathematics and their applications – quickly, informally and at a high level. Mathematical texts analysing new developments in modelling and numerical simulation are welcome.

   Manuscripts should be reasonably self-contained and rounded off. Thus they may, and often will, present not only results of the author but also related work by other people. They may be based on specialised lecture courses. Furthermore, the manuscripts should provide sufficient motivation, examples and applications. This clearly distinguishes Lecture Notes from journal articles or technical reports which normally are very concise. Articles intended for a journal but too long to be accepted by most journals, usually do not have this "lecture notes" character. For similar reasons it is unusual for doctoral theses to be accepted for the Lecture Notes series, though habilitation theses may be appropriate.

2. Besides monographs, multi-author manuscripts resulting from SUMMER SCHOOLS or similar INTENSIVE COURSES are welcome, provided their objective was held to present an active mathematical topic to an audience at the beginning or intermediate graduate level (a list of participants should be provided).

   The resulting manuscript should not be just a collection of course notes, but should require advance planning and coordination among the main lecturers. The subject matter should dictate the structure of the book. This structure should be motivated and explained in a scientific introduction, and the notation, references, index and formulation of results should be, if possible, unified by the editors. Each contribution should have an abstract and an introduction referring to the other contributions. In other words, more preparatory work must go into a multi-authored volume than simply assembling a disparate collection of papers, communicated at the event.

3. Manuscripts should be submitted either online at www.editorialmanager.com/lnm to Springer's mathematics editorial in Heidelberg, or electronically to one of the series editors. Authors should be aware that incomplete or insufficiently close-to-final manuscripts almost always result in longer refereeing times and nevertheless unclear referees' recommendations, making further refereeing of a final draft necessary. The strict minimum amount of material that will be considered should include a detailed outline describing the planned contents of each chapter, a bibliography and several sample chapters. Parallel submission of a manuscript to another publisher while under consideration for LNM is not acceptable and can lead to rejection.

4. In general, **monographs** will be sent out to at least 2 external referees for evaluation.

   A final decision to publish can be made only on the basis of the complete manuscript, however a refereeing process leading to a preliminary decision can be based on a pre-final or incomplete manuscript.

   Volume Editors of **multi-author works** are expected to arrange for the refereeing, to the usual scientific standards, of the individual contributions. If the resulting reports can be forwarded to the LNM Editorial Board, this is very helpful. If no reports are forwarded or if other questions remain unclear in respect of homogeneity etc, the series editors may wish to consult external referees for an overall evaluation of the volume.

5. Manuscripts should in general be submitted in English. Final manuscripts should contain at least 100 pages of mathematical text and should always include

   - a table of contents;
   - an informative introduction, with adequate motivation and perhaps some historical remarks: it should be accessible to a reader not intimately familiar with the topic treated;
   - a subject index: as a rule this is genuinely helpful for the reader.
   - For evaluation purposes, manuscripts should be submitted as pdf files.

6. Careful preparation of the manuscripts will help keep production time short besides ensuring satisfactory appearance of the finished book in print and online. After acceptance of the manuscript authors will be asked to prepare the final LaTeX source files (see LaTeX templates online: https://www.springer.com/gb/authors-editors/book-authors-editors/manuscriptpreparation/5636) plus the corresponding pdf- or zipped ps-file. The LaTeX source files are essential for producing the full-text online version of the book, see http://link.springer.com/bookseries/304 for the existing online volumes of LNM). The technical production of a Lecture Notes volume takes approximately 12 weeks. Additional instructions, if necessary, are available on request from lnm@springer.com.

7. Authors receive a total of 30 free copies of their volume and free access to their book on SpringerLink, but no royalties. They are entitled to a discount of 33.3 % on the price of Springer books purchased for their personal use, if ordering directly from Springer.

8. Commitment to publish is made by a *Publishing Agreement*; contributing authors of multiauthor books are requested to sign a *Consent to Publish form*. Springer-Verlag registers the copyright for each volume. Authors are free to reuse material contained in their LNM volumes in later publications: a brief written (or e-mail) request for formal permission is sufficient.

Addresses:

Professor Jean-Michel Morel, CMLA, École Normale Supérieure de Cachan, France
E-mail: moreljeanmichel@gmail.com

Professor Bernard Teissier, Equipe Géométrie et Dynamique,
Institut de Mathématiques de Jussieu – Paris Rive Gauche, Paris, France
E-mail: bernard.teissier@imj-prg.fr

Springer: Ute McCrory, Mathematics, Heidelberg, Germany,
E-mail: lnm@springer.com

Printed in the United States
by Baker & Taylor Publisher Services